629.13 R54a 2221469
Robertson, Bruce.
Aviation enthusiasts' data book

There is no mistaking the type name of this Grumman XF9F-2 fighter as Panther.

AVIATION ENTHUSIASTS' DATA BOOK

BRUCE ROBERTSON

Beaufort Books, Inc.
New York

Library of Congress Cataloging in Publication Data

Robertson, Bruce.
Aviation Enthusiasts' Data Book.

1. Aeronautics—handbooks, manuals, etc. I. Title.
TL551.R63 1984 629.13 83-22518
ISBN 0-8253-0208-0 (pbk.)

Published in the United States by Beaufort Books, Inc., New York.

Printed in Great Britain
First American edition, 1984

10 9 8 7 6 5 4 3 2 1

Contents

Introduction

This is a book for air enthusiasts. It aims to set the scene for those new to the appreciation of the world of aerospace, and to fill in the gaps for the older hands. However much we know, there are sometimes gaps in our knowledge or points which require clarification. Often during our general education missing out on particular lessons has left a blank, perhaps on some point of English or an arithmetical process; also we may have forgotten. It is difficult to find a comprehensive book of basics for our general education; indeed most of us would be reticent about being seen with a primer. But for the aerospace world this book is aimed to be just that, to provide general reference for the newcomer and a refresher for the mature enthusiast.

Books and magazines must assume that we have a breadth of knowledge, to avoid endless repetition. They often make us aware of our lack of grounding. Here is an attempt to give 'roots' to help an understanding, and thereby greater enjoyment, of the intriguing world of aerospace. In this task I have been greatly aided by Peter G. Cooksley, an author in his own right as well as an artist, who has charted in simple form many of the complex aspects of the subject. I am also indebted to E.F. Cheesman for his helpful suggestions.

Bruce Robertson
London 1981

Chapter 1

The world scene and aeronautics

Journalists and commentators in newspapers, on radio and television, usually assume that those addressed have a knowledge of current affairs and recent history which, indeed, we all do have to some extent. But not all are fully cognizant of just what is involved in, say, NATO air forces, for not all western European countries are NATO members—and two which are do not have air forces. Also, those with historical interests are sometimes a little bewildered that Allies of one war, are enemies in the next. So here, set out briefly, is the line-up of world powers and their alliances, followed by a background on the World Wars which have had so much influence on the development of aviation in particular.

The more your interest in aircraft grows, the more you will learn of recent history and current affairs. Do not take all you read or hear as fact. As you gain experience, you will be surprised at some of the misreporting by newspapers, radio and TV commentators on matters of aviation, and how often the picture shown on TV is not really correctly related to the commentary.

United and dis-united nations

The highest world authority is the United Nations Organisation (UNO), and practically every country of the world is a member. One of its main sections is the International Civil Aviation Organisation (ICAO) which controls worldwide the rules and regulations for air navigation. Aircraft of the armed forces of member nations do not come under ICAO, but are bound by international law as decreed by the International Court of Justice at The Hague, another body of the United Nations. In the case of aggression by nations, it was considered that, as it would take too much time for a full assembly of all United Nations' representatives to be convened, so a permanent Security Council, consisting of the five major world powers and six rotational other member nations, could meet and quickly resolve such matters. But owing to the original ruling that a major power could veto action, the Council has not always been effective. As a result, countries have sought their own alliances—the Western World accepting democracy, and many of the countries of the Eastern World adopting communism. In the Western World the United States of America is the major power by size and military strength and in the Eastern World there is the Union of Soviet Socialist Republics (USSR—often more briefly called the Soviet Union). The term 'Russia' is also loosely used, and it is proper to use this word in particular for aeronautics in that country up to 1918, as the socialist state did not come into being until after their October 1917 revolution.

What is the form? A wide range of configurations posed by US Navy aircraft at Patuxent River, Maryland, in the '60s (above) and by US Air Force aircraft at Eglin, Florida, in 1957 to celebrate the 50th anniversary of American military aviation (below). Here are seen high-wing, mid-wing and low wing monoplanes, conventional wings, swept wings and delta wings. There are also fixed wings, rotary wings and folding wings. Aircraft with two, four and six engines can be seen, including one, B-36 (below, top centre), with pusher engines. The aircraft scene is one of infinite variety.

Western alliance

Following the Second World War, Britain and France pledged mutual support in any future struggle under the 1946 Treaty of Dunkirk. The following year this Treaty was expanded as Western Union to include the Benelux countries (Belgium, Netherlands and Luxembourg). America had in mind a greater alliance of western powers and so, in 1949, the North Atlantic Treaty Organisation (NATO) came into being with America and Canada joining the Western Union countries and including Italy, Denmark, Norway, Portugal and Iceland. While the last-named and Luxembourg did not have air forces, their territory was of military importance to the alliance. In 1952 Greece and Turkey became members of the NATO alliance and West Germany joined in 1955. A large proportion of the British Forces and the forces of other member nations are controlled by NATO from its main headquarters near Brussels. Two complications in recent years have been the withdrawal of French forces from allocation (but France remains associated with the alliance), and the dispute between Greece and Turkey over Cyprus, which has a Greek and Turkish population. It is here that the RAF has its only remaining outpost in the Near East, in support of NATO, by agreement with the Cyprus government.

In Western Europe, those countries left outside the NATO alliance are the traditionally non-aligned countries, Switzerland and Sweden. Although the latter is socialist to the extent of recently dropping the 'Royal' prefix from its Air Force, it is opposed to communism and, like Switzerland, is strong and well organised for self defence. Austria and Finland, re-established as independent nations by treaty agreements between East and West, are also non-aligned. Spain was not asked to join NATO because of the fascist image of Franco, but since the death of that dictator the country has become a candidate for membership. Disregarding principalities like Liechtenstein, the only other West European country not involved in NATO is Eire.

Eastern alliance

In Eastern Europe the Soviet Union has been instrumental in arranging a military alliance of communist countries under the Warsaw Pact, embracing the USSR, Bulgaria, Czechoslovakia, East Germany, Hungary, Poland and Romania. The USSR's allies are often referred to as the Soviet satellites. Albania, once a satellite and in the Warsaw Pact, has since opted out. China and Yugoslavia, both communist countries, have no military commitments to other communist countries.

SALT

It is to be hoped that a Strategic Arms Limitation Treaty (SALT) will be able to limit arms as a step towards world peace. However, this requires goodwill on the part of East and West and unless that goodwill can be guaranteed, by mutual inspection (which has not yet been fully agreed) then, with both factions fully armed, it is the fear of the opponent's strength which is the main deterrent to war. This is not a satisfactory way to ensure peace, but while it is the only way it must be seen to be effective or fail in its purpose. For 35 years, the deterrent, the threat of retaliation by nuclear missiles, has kept the peace between the major powers, in the same way that gas, the horror weapon of the First World War, stayed the unleashing of chemical warfare in the Second World War. This is the state of the world today, in which aviation plays a vital part.

Background

The 1914-18 War was known as The Great War until 1939 when the Second World War caused its renaming as the First World War. Officially, by governmental decree, it was The Great War of 1914-19. Although hostilities ceased on the Western Front on the eleventh hour of the eleventh day of the eleventh month of 1918, fighting went on into 1919 in the north and south of Russia where Allied troops with RAF support were aiding the White Russian forces resisting the 'Red' revolution which founded the USSR. Also numbers of men died of disease contracted on active service and the immediate post-war tasks, such as mine clearing, took their toll; hence many war memorials of that period bear the inscription 1914-19.

The war started late in July 1914 with Austro-Hungary attacking Serbia. Russia aided Serbia while Germany sided with the old Austrian empire. Whilst fighting Russia on one front Germany also invaded France and violated Belgian territory, thus provoking a declaration of war from Britain on August 4. The Germans and Austrians soon had the support of Turkey and then Bulgaria. France and Britain, Belgium's allies, held the Germans on the Western Front which stretched from the Belgian coast to the Swiss border. Britain, then with a vast empire, soon had the support of Australia, Canada, India, New Zealand, South Africa and a host of smaller colonies. Japan joined the Allies later in 1914, contributing mainly naval forces. Italy joined in 1915 engaging the Austrians, then Romania joined the Allies and finally the USA in April 1917. Apart from the Western Front the British were involved in aerial activity in: the Mediterranean and Egypt, Palestine and Mesopotamia (then under the old Turkish Empire), the Dardenelles 1915-16 (in an abortive attempt to control the straits), in Salonika and Italy 1916-18 and in parts of Russia 1918-20.

Post-war, the British Government had a mandate from the League of Nations, the unsuccessful forerunner of UNO, for policing Mesopotamia (renamed Iraq in 1921), which was carried out by the Royal Air Force, and the unhappy burden of policing Palestine carried out with the co-operation of the RAF. With the British also having responsibility for the peace of India, the RAF and troops were frequently involved on the North-West Frontier with Afghanistan, preventing the fierce hill tribes from plundering the plains. This area is now in Pakistan as a result of the partition of India when the British left in 1947. The RAF did not take a hand in the defences of the Far East until the '20s when Singapore and Hong Kong became the main air bases, the latter now being the only base with an RAF elcment permanently in the Far East.

The Second World War had its beginnings in the rise of Nazi Germany under Hitler and a partnership with Italy, then a fascist country, under Mussolini. They became the so-called Axis Powers, expanding their influence from a Berlin-Rome axis. Italy with colonies along the North African and Red Sea Coast, bit deeper into Africa by conquering Abyssinia (as Ethiopia was then known) during 1935-6 and later invaded Albania. Hitler meanwhile had annexed Austria and, when Czechoslovakia was menaced similarly in September 1938, it led to the Munich crisis and the hasty camouflaging of all RAF aircraft remaining in peace-time 'silver' finish. War was averted, but when on September 1 1939 Germany attacked Poland, France and Britain declared war on Germany who, this time to avoid a war on two fronts, had signed a non-aggression pact with Russia.

As before, there was a Western Front from the Channel coast to the Swiss border with British troops (backed by the RAF) again at the north-western end of the line, near to the sea. Once more the Empire responded, Canada's declaration of war actually preceding our own, but Eire, which until 1922 had been part of the United Kingdom, remained neutral. In April 1940 the Germans invaded and occupied Norway and Denmark, which had remained neutral in the First World War. That May they swept through Holland (also previously neutral), Belgium and France, causing the evacuation of the British Expeditionary Force, isolating Britain from the continent, the eastern seaboard of which, from the north of Norway to the Bay of Biscay, then fell into enemy hands.

The Battle of Britain in the autumn of 1940 was then inevitable, as was the blitz which followed. However, this critical situation eased in 1941 as German commitments grew when Hitler joined Italy in an attack on Greece and then declared war on Yugoslavia that April. The Germans also joined Italian forces in North Africa, invaded Crete with airborne forces in May and then, on June 22, invaded Russia. That same month Italy, Hungary and Romania declared war on Russia, as did Finland which had suffered earlier by an attack from Russia, and so became a technical enemy of the Allies. Britain and Russia, as allies, jointly occupied Iran to guard oil supplies in the Middle East which the Germans sought to reach by a pincer movement through the Balkans and along the North African coast to Egypt.

Meanwhile in the Far East, the Japanese, at war with China since 1931 and occupying great tracts of that country, had joined the Axis in 1940, but not until December 7 1941 did they strike, launching attacks on US military, naval and air bases in Hawaii, including Pearl Harbor. The following day a Japanese force landed in Malaya and swept down the Malayan archipelago capturing Singapore and the Dutch East Indies which, in post-war years, became Indonesia. Moving westwards through Siam (renamed Thailand on May 11 1949) and from China, the Japanese over-ran Burma and threatened India. Bases in India and in Ceylon (now Sri Lanka) came under air attack, while Hong Kong fell to the Japanese that Christmas.

America, sympathetic to Britain who had been buying aircraft in bulk from the USA since 1938, passed the Lend/Lease Act in 1941 saving Britain massive payments for aircraft. The Pearl Harbor attack brought America immediately into the war with a policy for defeating Hitler first and then dealing with the Japanese. On May 12 1942, the first contingent of the US 8th Air Force arrived in Britain, the vanguard of a day bomber force which would join the RAF night bombers in a round-the-clock attack on Germany. As evidence of Britain's own growing strength, the first of the RAF's 1,000-bomber raids was on the night of May 30/31 1942 when 2,000 tons of bombs were dropped on Cologne in 90 minutes by 1,048 aircraft, of which 44 were lost. That August, Brazil declared war on the Axis and later sent an operational squadron to Europe.

In November 1942, an Allied force landed in North Africa, sandwiching the German and Italian forces engaged in battle with British and Commonwealth troops operating from Egyptian bases. The Axis troops having been driven out of Africa, and Sicily and then southern Italy having been invaded, the Italians themselves disposed of Mussolini and declared themselves on the side of the Allies. But there was still a hard slog up through Italy against German and some pro-fascist Italian troops. The Allied invasion of the continent was made from

Without its RAF roundels Westland Whirlwind HAR10 XK970 patrols in Cyprus on behalf of the United Nations as its UN marking indicates.

Britain in 1944—Normandy being attacked on June 6—(this was called Operation Overlord and the air plan was Operation Neptune). There were also Allied landings in southern France on August 17. In an abortive effort to shorten the war, the Arnhem airborne attack was carried out from September 17. More successful was the airborne operation in support of the crossing of the Lower Rhine near Wesel on March 24 1945 involving some 8,000 aeroplane and 1,300 glider sorties. The German unconditional surrender to the Western Allies and the USSR was given on May 7 1945 with the following day of celebration being called VE Day (Victory in Europe Day).

The re-conquest of Burma had already started with an air and amphibious attack on Rangoon. In the Pacific the Americans had been 'island-hopping' towards Japan and building bases for air attack on Japan in India and China as well as the Pacific islands. After a warning was given to the Japanese government, the first atomic bomb was dropped from a Boeing B-29 on Hiroshima, on August 6 1945 and three days later another was dropped on Nagasaki. Between those two events, Russia declared war on Japan who surrendered unconditionally on August 14 and two days later VJ Day (Victory over Japan Day) was celebrated.

Setting the aeronautical scene

The chart opposite shows the range and scope of aeronautics. Some of the terms originated from before the turn of the century when 'aerostats' were the only successful aircraft, but this term for lighter-than-air craft, and the use of 'aerodyne' for heavier-than-air craft, have died out—although aerodyne is the

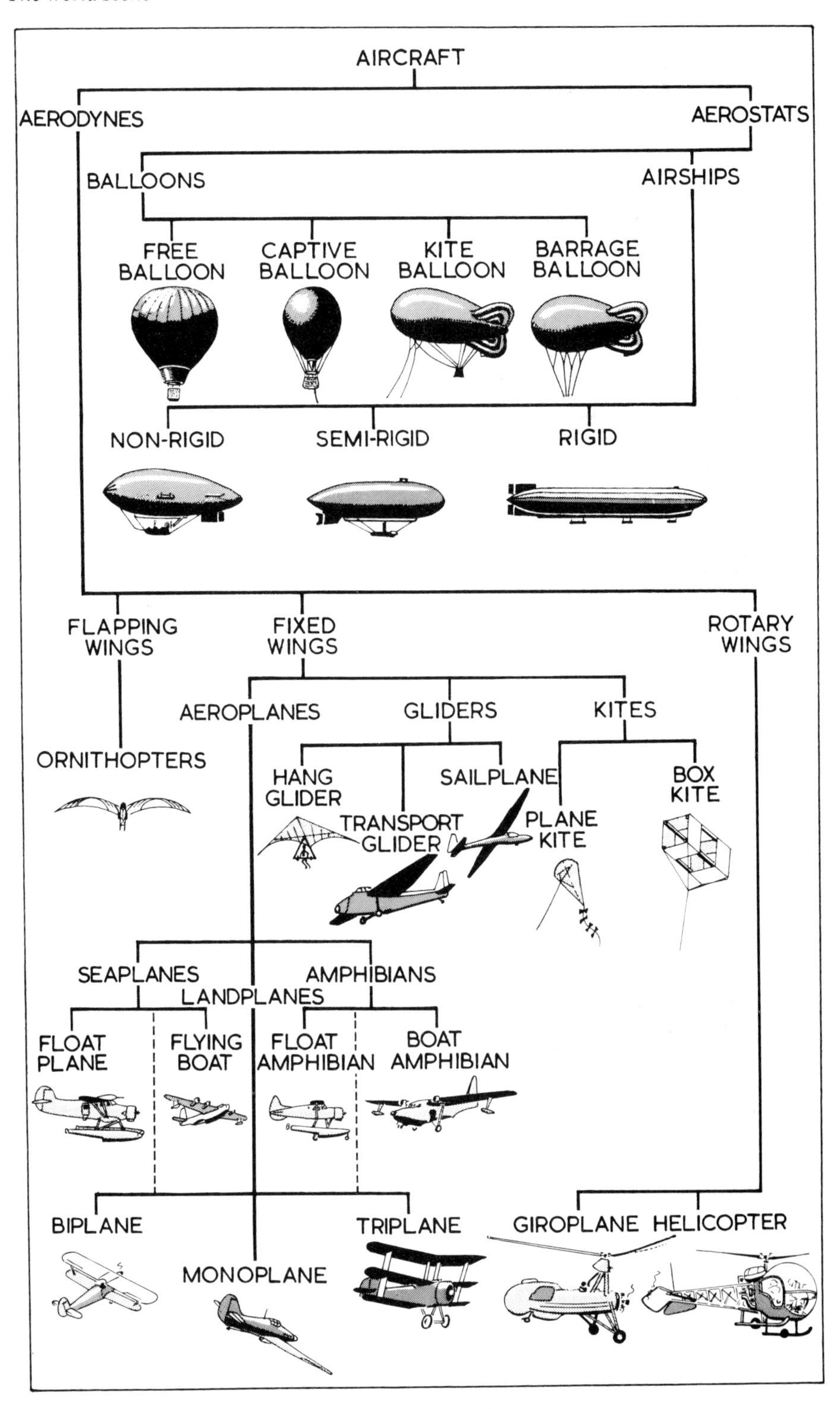
AIRCRAFT
AERODYNES
AEROSTATS
BALLOONS
AIRSHIPS
FREE BALLOON
CAPTIVE BALLOON
KITE BALLOON
BARRAGE BALLOON
NON-RIGID
SEMI-RIGID
RIGID
FLAPPING WINGS
FIXED WINGS
ROTARY WINGS
AEROPLANES
GLIDERS
KITES
ORNITHOPTERS
HANG GLIDER
SAILPLANE
BOX KITE
TRANSPORT GLIDER
PLANE KITE
SEAPLANES
AMPHIBIANS
LANDPLANES
FLOAT PLANE
FLYING BOAT
FLOAT AMPHIBIAN
BOAT AMPHIBIAN
BIPLANE
MONOPLANE
TRIPLANE
GIROPLANE
HELICOPTER

A post-war boat amphibian, the Piaggio P-136-L of Italian design was also built in America under licence as the Royal Gull.

only word to describe those paper darts of schooldays.

Aeronautics embraces all man-built objects flying within the stratosphere. Outside this realm we come into astronautics which is concerned with spacecraft and satellites. Rockets, although they have been used for mail-carrying, do not come into the category of aircraft, being allied to projectiles. The main portion of aeronatutics is now 'aviation', which refers to heavier-than-air machines and is a term often used now instead of aeronautics, but it is important to remember its limitations when talking to hot-air ballooning enthusiasts!

It is usual to call gyroplanes 'autogyros' (most dictionaries accept giroplane and autogiro as alternatives), but the latter was actually a tradename for a type of giroplane.

With the limited number of seaplanes and amphibians, it has been usual to call landplanes aeroplanes and so rid ourselves of the anomaly of calling a carrier-borne aeroplane a landplane, which it is by configuration. The Royal Naval Air Service used the term ship-borne aeroplane but the term fell into disuse in the early days of the RAF.

Perhaps the most controversial word is seaplane, embracing flying-boats and floatplanes. As originally intended, 'seaplane' covered all aircraft operating from water; the Services certainly took seaplane to mean all such aircraft. When, in the '20s, the flying-boat superseded most of the floatplanes in the RAF, the term flying-boat came into more general use and floatplanes came to be called seaplanes; but, correctly, seaplane covers all types and some dictionaries have been misled on this point.

Chapter 2

The language of aeronautics

English is one of the two languages officially recognised by the International Civil Aircraft Organisation (ICAO) which has its headquarters in Montreal. Canada is an appropriate country to house the ICAO as it is a dual English/ French speaking country and French is the other main internationally recognised language of the air. In the military sphere, in NATO (North Atlantic Treaty Organisation), the two official languages are also English and French. *Jane's All the World's Aircraft*, the yearbook internationally recognised as the premier aircraft reference and supplied worldwide, is printed only in English because this is the prime language of aeronautics.

Although the official language of the United States is English, we know well enough that they have their own way of spelling, pronouncing and inventing their own words. Some of their terms are very apt, but on the other hand a writing cult has developed complicating their writing by long and unnecessary wordage called Americanese. A very simple example of this is aircraft data on speed, range, etc, which would be called 'performance' in English and 'performance characteristics' in Americanese. Unfortunately it has had some effect on British writers. But look at it this way—if you cannot understand some writings, do not blame your own intelligence; regard it as the lack of ability of the writer to communicate.

First let us decide what we mean by 'aircraft'. The first chart shows that the term has a wide meaning and embraces the many forms aircraft can take. Because so much aeronautical development has resulted from the threat and pressure of war, service usage has had a great influence on aeronautical terms. The most widely used type of aircraft is the aeroplane, which Americans call an airplane. Yet this specific word is rarely used, and the very broad term aircraft is used in its place. This is because the RAF, in the mid-'20s, when it had only aeroplanes to fly, insisted that the service term for aeroplane would be aircraft. Now, with helicopters in service which are also aircraft, the Service uses the clumsy terms 'fixed wing' and 'rotary wing' to differentiate between aeroplanes and helicopters—and so we, too, continue to use the broad term aircraft for aeroplanes.

Since the French once led the world in aviation, many of the terms relating to aeroplane structures in particular have their origins in French. The body of an aircraft is a fuselage from the French *fusel* for shuttle—the body of aeroplanes being shuttle-shaped. Fuselages were originally built of wood, fabric-covered and doped to make the linen taut. The four main longitudinal (lengthwise) members of the box-like structure were the *longerons*—another French word.

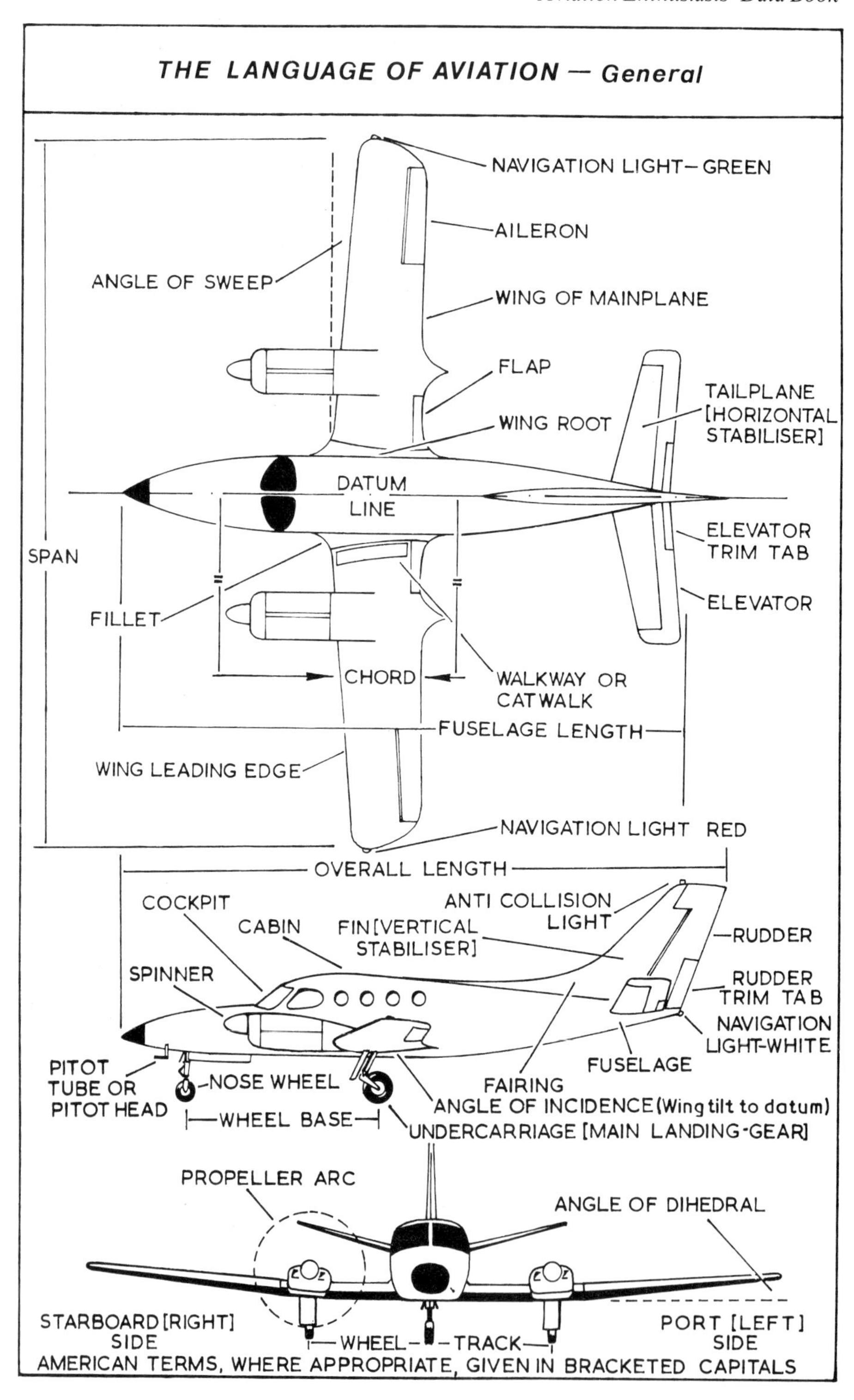
THE LANGUAGE OF AVIATION — General
NAVIGATION LIGHT—GREEN
AILERON
ANGLE OF SWEEP
WING OF MAINPLANE
FLAP
TAILPLANE [HORIZONTAL STABILISER]
WING ROOT
DATUM LINE
SPAN
ELEVATOR TRIM TAB
ELEVATOR
FILLET
CHORD
WALKWAY OR CATWALK
FUSELAGE LENGTH
WING LEADING EDGE
NAVIGATION LIGHT RED
OVERALL LENGTH
COCKPIT
ANTI COLLISION LIGHT
CABIN
FIN [VERTICAL STABILISER]
RUDDER
SPINNER
RUDDER TRIM TAB
NAVIGATION LIGHT-WHITE
FUSELAGE
PITOT TUBE OR PITOT HEAD
NOSE WHEEL
FAIRING
ANGLE OF INCIDENCE (Wing tilt to datum)
WHEEL BASE
UNDERCARRIAGE [MAIN LANDING-GEAR]
PROPELLER ARC
ANGLE OF DIHEDRAL
STARBOARD [RIGHT] SIDE
PORT [LEFT] SIDE
WHEEL—TRACK
AMERICAN TERMS, WHERE APPROPRIATE, GIVEN IN BRACKETED CAPITALS

With modern aircraft the construction technique has entirely changed with the fuselage's strength being in the covering. Sections are used to form a shell or *monocoque*, meaning single hull.

A standard dictionary is not always helpful in assisting with aeronautical terms. It is easy enough to explain that a monoplane has one set of wings (planes) and a biplane has two sets, but one dictionary gives: 'Monoplane—A flying machine with its wings or carrying surfaces arranged in nearly the same plane'. The Germans call the wings our equivalent of decks, so a monoplane is an Eindekker, a biplane a Doppeldecker and a triplane a Dreidekker.

The Americans have their own words for parts of an aircraft structure, for example 'horizontal stabiliser' for tailplane. A chart has therefore been prepared to show where there are differences between American and English terms.

Most early aircraft had the engine behind the wings. When aircraft appeared with the engine in front, it led to them being called tractor aircraft, because their propellers pulled them through the air. When they became the normal configuration for aircraft during the First World War, the word tractor was dropped and it was the aircraft with the propeller in the rear which were then called pusher aircraft and are so-called to this day. The 'control lever' of the past, the 'joystick' in slang, officially became the 'control column' in RAF usage from 1919 and this was generally adopted. It was at the same time that the RAF decreed that the term 'undercarriage' would be used in place of chassis or landing gear.

Pilot has become the usual word for the person actually responsible for flying the aircraft and he, under British rules of the air, is the captain of a passenger aircraft and first pilot if more than one pilot is carried. In a Service aircraft the flying crew are collectively called 'aircrew' and in a passenger aircraft crews are divided into flight-deck crew (pilots, navigators, radio-operators) and cabin crew (air hostesses, bar-tender, etc). In the past airship crews were known as aeronauts and aeroplane crews as aviators; these words have now practically disappeared.

Up to the middle of the Second World War a flying field with hangars and other facilities was called an 'aerodrome', but this was then changed to 'airfield' to be more in line with our American allies. An airfield used for international flights with a Customs unit has the status of an airport.

Air operations, aero-engines, armament and radio all have a language of their own and these are dealt with in later chapters. This introduction to the language of aeronautics must of necessity be brief for, to explain all the terms, needs a very large aeronautical dictionary indeed.

Chapter 3

For good measure

Facts and figures play a large part in aviation, providing a means of assessing comparative merits, expressing capabilities and giving an idea of size. But first let us be sure about our sense of values. The figures normally given for dimensions and performance do *not* record the vital factors. For what do we seek in a good aircraft? Most commercial operators look for four qualities, in this order—safety, reliability, economy and maintainability. To an extent these are criteria for military aircraft, too, but in addition manoeuvreability is also a critical factor—it certainly was in the two World Wars. Yet just where do you find the comparative figures to measure these qualities? The answer is that you do not.

So we have to content ourselves with data which can easily be expressed in facts and figures. The order in which these are presented can vary; there are no set rules. But it is a strange thing that with military aircraft, practically the last factor mentioned is armament. Yet the whole object of these aircraft is to enable the weapon to be brought to bear. A bomber or fighter is a powered weapon platform, yet the weapon usually comes low in the listing of its characteristics.

The figures which are given first should be those relating to the human element—the number of crew and persons carried; with small aircraft, single-seat or two-seat, these words normally come into the basic description. With two-seat aircraft it is usual to specify if the seats are side-by-side or tandem (one behind the other). With tandem seating, it should not be assumed that the pilot's seat is invariably in the front.

In a large aircraft crew numbers may vary considerably. Basic operations crew may include a pilot (captain), co-pilot and radio-operator; then there are the cabin staff. A Boeing E-3A Sentry AWACS aircraft, for example, has an operating crew of four and a specialist crew of 13. Accommodation usually relates to passengers, but a so-called two- to six-seat light aircraft, normally includes the pilot. On larger aircraft the accommodation can vary greatly even in a single type of airliner according to whether it is fitted out for VIP transit, scheduled route with different classes, or for tourists.

Measures—metric and Imperial

Before we can look at the measurements of basic data there is a complication. At the present time, we have two main systems of measurement to contend with—Imperial units (feet, pounds, etc), which have been our traditional measures for well over a century and the metric units (metres, grams, litres, etc) which are used on the continent.

The metric system is no stranger to the aeronautical industry and some aircraft firms have always dealt in metric units; this dates from the early years of the century when France led the world in aviation and in the First World War Britain was largely dependent on French aero engines, which were built here under licence. However, most of us have been brought up on the Imperial system and many of our reference books use only these units. Nevertheless, we will all have to get used to metric units as it has been government policy since 1961 to adopt the International Metric System. As originally planned it was to be phased in by set dates, but the introduction of some units has been delayed. As far as can be foreseen we are committed to metric. The phasing in can already be seen in such examples the replacing of Fahrenheit by Celsius temperatures on television weather forecasts. The introduction is usually made by having a period in which the new system appears in parentheses, then there is a change-over to the new system with the old in parentheses, and finally the old system is discarded.

The basis of the metric system is water, as will be explained. For practical purposes in metric, the metre (abbreviated to m and equivalent to 3.3 feet) is the basic unit for aircraft overall measurements. A thousandth part of a metre is a millimetre (mm), the smallest measure in the system and is used in particular for classifying the calibre of weapons, as will be explained under 'Armament'. Ten millimetres (or a hundredth of a metre) is a centimetre (cm), more generally used in component manufacture than relating to complete aircraft. But a cubic centimetre provides the link between linear measures and weight and volume in the metric system, for the basic unit of weight, the gram, is the weight of a cube of water with sides of 1 centimetre. 1,000 grams is 1 litre by volume and 1 kilogram by weight (2.2 lb in Imperial measure). Going back to linear measure to complete the metric scene, 1,000 metres are 1 kilometre, the unit roughly equivalent to $\frac{5}{8}$ of a mile, used for expressing distance.

With its basis as water, the system was admirable in the days of steam power, when it came into general use in France in the middle of the last century. However, the weights of fuel and oil in our aeronautical age are not those of water.

Jane's All the World's Aircraft which had always given metric measures, but in parentheses following Imperial measures—now leads with metric. It is a sign of the times. The officially recognised authority for promulgating records of airships and aeroplanes over 70 years, and helicopters since 1924, is the Fédération Aéronautique Internationale, based in Paris, so that official aircraft record figures are, and always have been, measured and recorded in metric. So, we have to go along with it in aeronautics, with only our measures of angles and time left unthreatened by metrication.

Basic data

The primary measurements of an aircraft are always given in metres or feet (never yards). Fractions of a foot may be expressed as a vulgar fraction or as a decimal—but in these days of impending metrication, it should be remembered that $\frac{1}{2}$ foot = 6 in, which is 0.5 ft not 0.6 ft—it is surprising the number of people who make this small error. Fractions of a metre must always be expressed in decimals. The primary measurements needed to indicate size are: wingspan, length and height for an aeroplane or glider; length of fuselage and diameter of rotor for a helicopter; and length, breadth (beam) and height for an airship.

Span is the measurement from wingtip to wingtip and refers to the longest wing in the case of a biplane. With multi-winged aircraft, where individual spans are recorded, it is usual to give the upper wing first, then the middle wing, in the case of a triplane, and finally the lower wing. For aircraft recognition training purposes, span is the only measurement given for an aeroplane and is normally to the nearest foot or half-metre, for in this case the figure is only to give an indication for anti-aircraft weapon setting.

Span has its complications. What was the wingspan of a Spitfire? Six different spans can be quoted with an 8 ft variation, according to mark, and also within the same mark according to whether stub wingtips were fitted for low altitude operations or extended tips to increase lift for high flying. In the case of aircraft today which have detachable wingtip fuel tanks, two spans can be quoted and, with swing-wing aircraft, the spans fully extended and fully swept are normally quoted.

Length is normally the overall length, not including whip-type aerials which might protrude forward of the nose. It is normally the only measurement given for helicopters for recognition training purposes, since the rotors, which appear so plainly in photographs taken at high shutter speeds, are rarely visible to the human eye during flight.

Height is the height standing, and not the height by the datum line (see the section on Plans) or in flying position, for the very good reason that this measurement is given for hangar ceiling height reference. In practice height would vary slightly according to aircraft loading, the figure, therefore, relates to the maximum height, with the aircraft unloaded. This measurement normally related to airframe height and did not include propeller or armament when this projected higher than the airframe.

Wing measurements, some expressed in angles, as illustrated in the charts, relate to the mainplanes, as they are also called. Should you be rusty in your appreciation of angles, then they are the measurement of the divergence of two lines, measured in degrees with 360 degrees to the full circle. To visualise the sweepback of a wing given at, say, 24 degrees, look at your watch. It is divided into 60 minutes which is a sixth of 360; so divide the angle by six ($24 \div 6 = 4$), thus the angle will be the same as that formed by your watch hands four minutes apart.

The angle of the slope of the wings (see drawing page 14) is dihedral, but should they droop, the angle is given with a minus sign prefix to denote a negative dihedral, or anhedral as this condition is called. The tilt of the wing to its line of flight, also illustrated, is the angle of incidence, but this, normally a low angle, is rarely ever negative.

With a biplane the distance between the wings is called the gap, but as it means the distance between the centre of the leading edge of each wing, it is approximately a wing thickness more than the true gap in the normal usage of the word. A biplane with a lower wing less than half the area of the upper wing, is called a sesquiplane.

Wing chord, the breadth of a wing, is less complicated with a straight-winged aircraft, but will vary along its whole length if the wing is tapered. The craft model builder will need to know the chord at a given distance along the wing, from the datum line—the centre of the aircraft. For non-straight winged

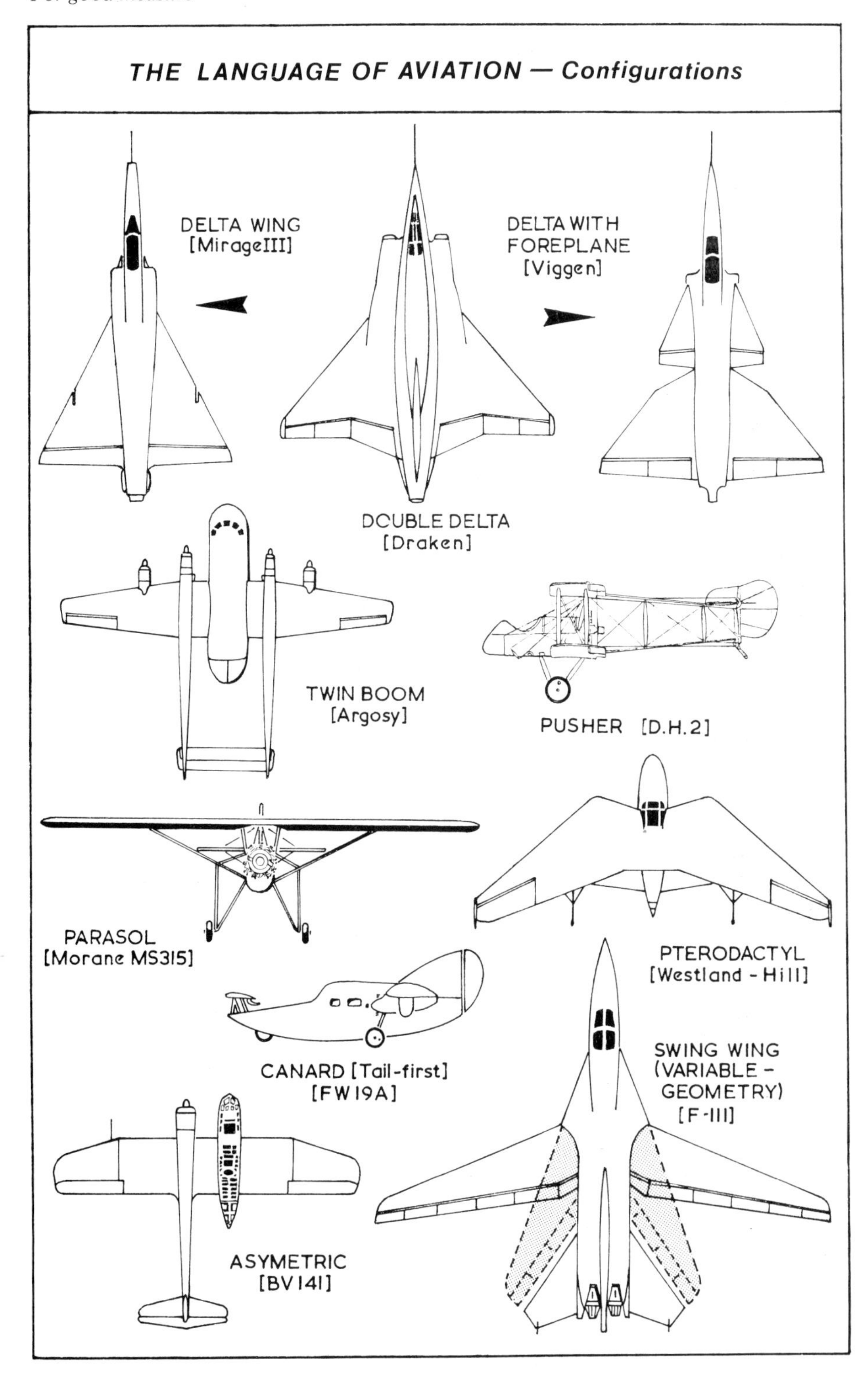
THE LANGUAGE OF AVIATION — Configurations
DELTA WING
[MirageIII]
DELTA WITH
FOREPLANE
[Viggen]
DOUBLE DELTA
[Draken]
TWIN BOOM
[Argosy]
PUSHER [D.H.2]
PARASOL
[Morane MS315]
PTERODACTYL
[Westland - Hill]
CANARD [Tail-first]
[FW19A]
SWING WING
(VARIABLE -
GEOMETRY)
[F-111]
ASYMETRIC
[BV141]

THE LANGUAGE OF AVIATION — Features

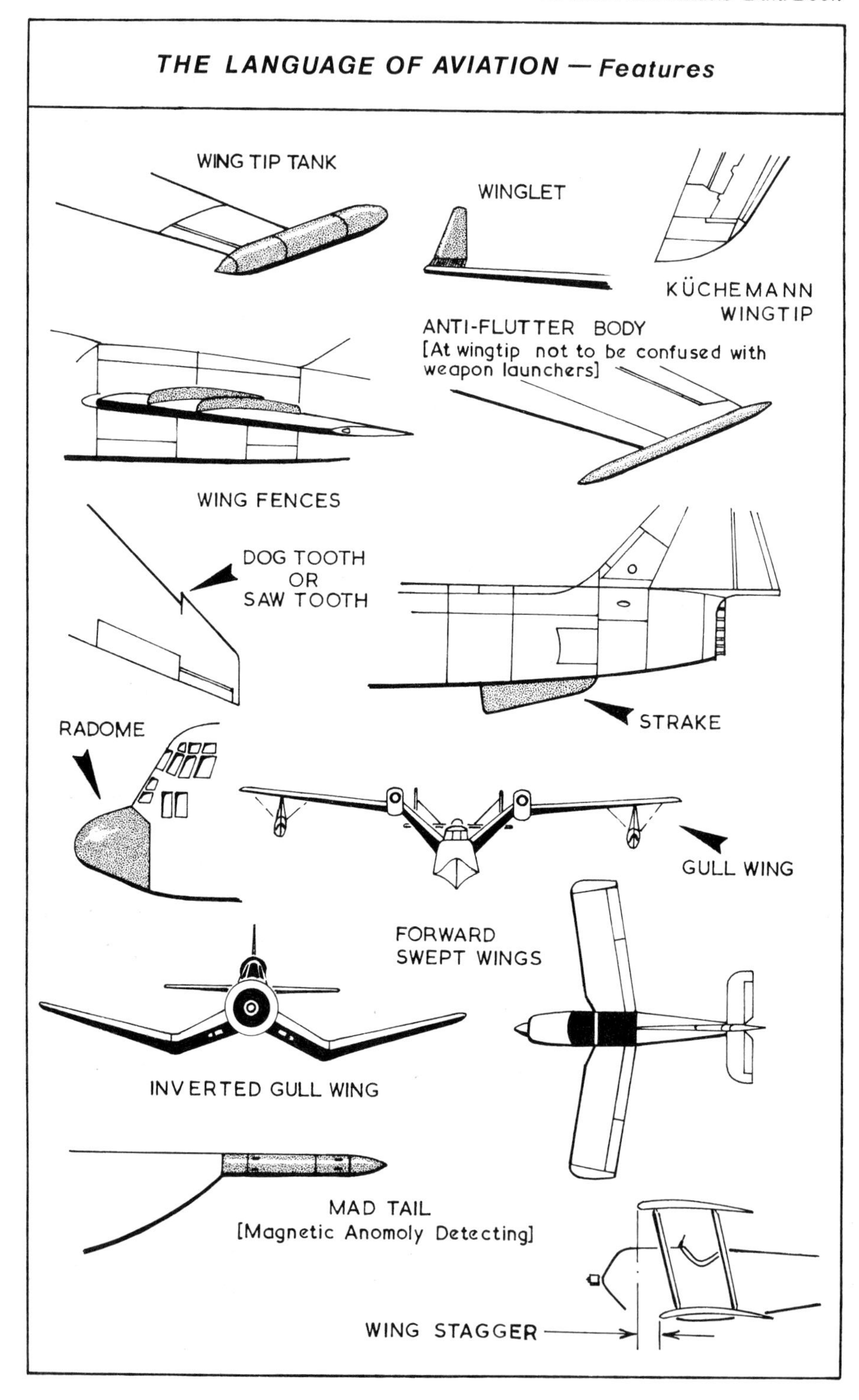

THE LANGUAGE OF AVIATION — Helicopters

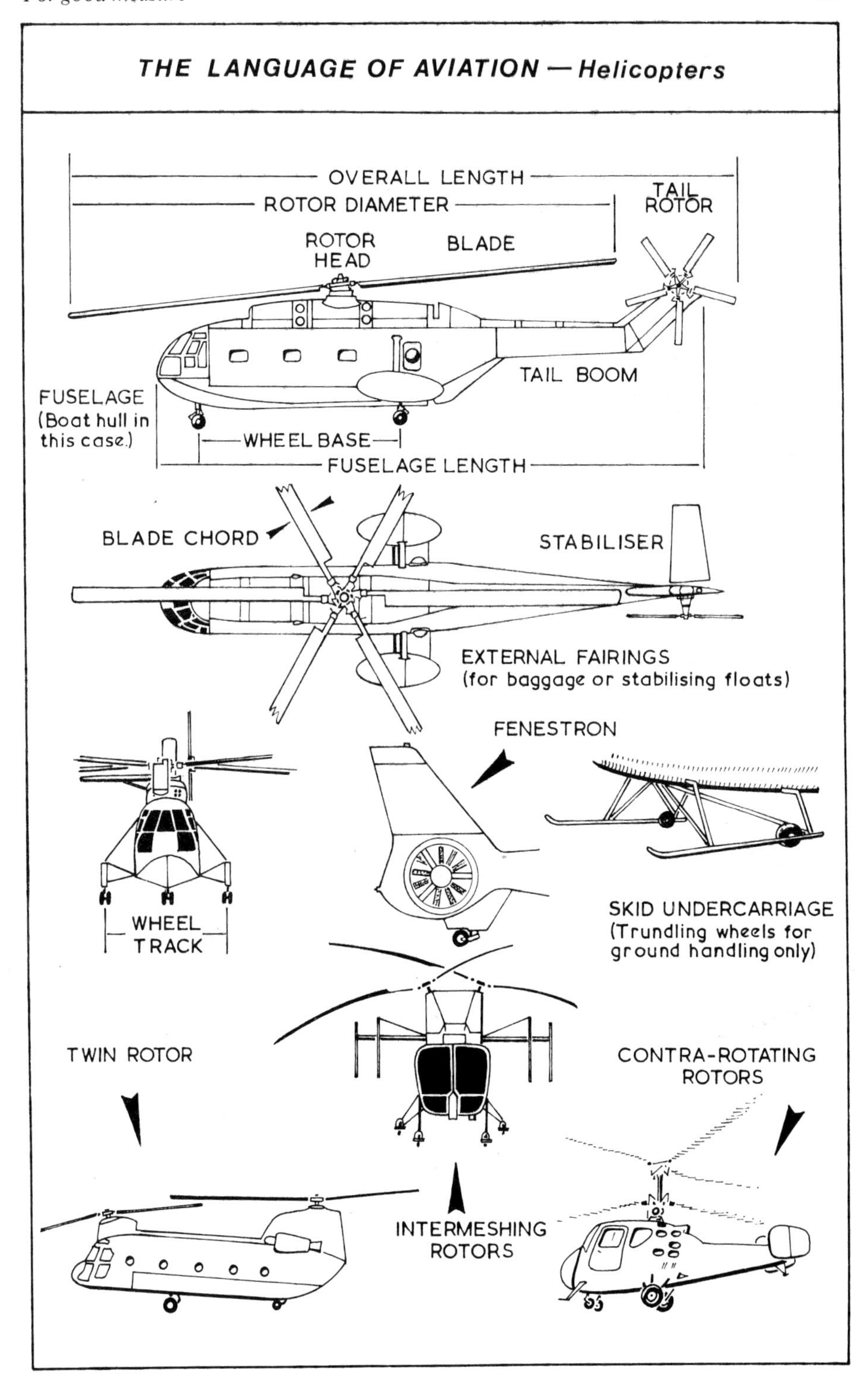

Wing stagger is the measurement of the sit of the wings in relation to each other. In most cases they are level or the bottom wing is placed slightly behind the top wing, but the DH5 fighter (top) had a negative stagger with the bottom wing in advance of the top. This was also a feature of the DH6 trainer (above) (Airco).

aircraft, the breadth of the wing may be expressed as a mean (average) chord. With wing length and average breadth known, their multiplication will give the wing area. This has always been considered an important figure, for the lift of an aircraft is proportional to the area of wing. Figures given for wing areas include ailerons and trimming tabs, but exclude flaps and/or slats. A significant figure of aircraft rating is its wing loading, expressed in kilograms per square metre or pounds per square foot of wing, eg, the Dakota was rated at 124.5 kg/m^2 or 25.5 lb/ft^2.

Aspect ratio is yet another figure quoted for wing shape. Aspect in this case means precisely what it implies—appearance of a wing whether long and narrow or short and stubby. The ratio is that of the length to the breadth of the wing, eg, a wing 60 ft long and 10 ft broad (chord) will have an aspect ratio of 60 ÷ 10 = 6. Meaning, in words, that its length is six times its breadth. It should be appreciated that a wing length is less than half span, as span includes fuselage width.

Wing section, or airfoil, is the streamlined shape which would be seen if the wing was cut in half and looked at end-on. This cannot be given by dimensions and is quoted by a design authority number, eg, Gottingen 398, RAF30 (RAF for Royal Aircraft Factory pre-1918), or NASA4409 (for National Aeronautical and Space Administration of USA). The shape will be decided by wind-tunnel tests. Some firms use their own sections and in some aircraft the section shape varies from root to tip. Typical sections are illustrated.

RAF15 *NACA4412* *SIKORSKY GS-1*

Weights have always been important figures in aircraft design and operation. At the top end of the weight scale we have tons, but a ton can be three different weights. There is the British measure of 2,240 lb, which is called a long ton to distinguish it from the American ton of 2,000 pounds called the short ton. But there is also a metric tonne, (note the different spelling which is abbreviated to just t) which is 1,000 kg (1 megagram) or 2,200 lb. However, tons are only dealt with when referring to bulk supplies such as in an airlift, or when an aircraft is spoken of loosely as able to carry so many tons of cargo or bombs.

The weight data for aircraft are dealt with only in kilograms or pounds, no matter how high the figures go, they are not broken down into units such as stones or hundredweight. The maximum weight of an aircraft is its ramp weight fully loaded and ready for take-off—a figure of importance to airfield construction. For flying, its maximum take-off gross weight (MTOGW) is its ramp weight less the weight of the run-up and taxying fuel; this will be rigidly specified for safety considerations and is the reason why air travellers' baggage is so meticulously weighed. There is also an all-up-weight (AUW) which is normally the total weight of MTOGW or below, when weights are set out for particular operating conditions, eg, for different cabin lay-outs. Subtract passengers, their baggage and/or cargo, which collectively is called the payload, and you have the basic operating weight (BOW). Remove crew and bar stocks, and empty the tanks to get rid of all disposable load and you have the bare weight, a figure more of importance for small aircraft when they are to be transported by land or sea. The weight of fuel is an important consideration in aircraft operation and this brings us to volume.

Volume in aviation concerns mainly the capacity of fuel and oil, collectively called tankage, and this is expressed in litres or gallons. Again we have a complication of British and American gallons—see the tables for values. Where capacity is given as internal, the inference is that auxiliary tanks can be carried externally. But where extra fuel is supplied by this means, and by using cabin space for tanks in ferrying, a corresponding rise in oil tankage has to be considered.

To most air travellers, the volumes which directly concern them, freed as they are from duty payments, are the bar stocks. Beer as yet is still in Imperial measure, but on American airlines a gill (quarter pint) is a nip, a half (half pint) is a small, and a pint is called a large. With spirits a tot can vary from a sixth to a third of a gill, but a noggin is a gill. For wine, you may get a bottle ($1\frac{1}{3}$ pints) with your meal, but for duty-free purchasing the tendency now is to deal only in litre bottles. Cheers!

In passing—and they did pass but are coming back in a limited way—airship capability is measured basically by its volume, since this represents the amount of gas which will provide lift. This capacity is given in cubic metres (m^3) or cubic feet (ft^3). Lift will also depend on the gas used; helium is more efficient and expensive than hydrogen, which was used in Zeppelins and led to several disasters.

Performance of an aircraft is dependent on the power available, so the basic data on any aircraft include its engine type name and power rating. Since engines are a vast subject in themselves, your briefing on engines is given separately. Most figures of performance are rather meaningless unless the condition of the aircraft is quoted. Speed, range, landing run, etc, are all dependent on the loading of the aircraft.

Speeds are all dependent on a variety of factors. For an interceptor, maximum speed is all-important; for an airliner it is cruising speed, at which it is most economical, which is all-important. But the cruising speed varies according to height, so speed is usually given as at a particular height. Manufacturers usually quote the best figure, while official reports tend to give speed at phased heights for comparative assessment purposes. The 10,000 ft point is a nice round figure but, as this is 3,050 m, it is not a convenient point at which to give continental figures. But there is yet another problem, the speed of an aircraft is given in knots, miles per hour (mph) or kilometres per hour (km/h).

Britain, the traditional seafaring nation, measured distance at sea in nautical miles and this was adopted for aerial navigation. A nautical mile is 6,080 ft compared to the 5,280 ft of the standard mile. To travel one nautical mile in one hour gives the speed of one knot. Speeds of aircraft have been recorded in knots from the earliest days of aviation and it is the measure of speed used by the RAF.

The foregoing measures are used for aircraft flying below the speed of sound, ie, subsonic. Travel faster than the speed of sound, ie, supersonic, or speeds around that range are quoted by Mach number. The name Mach comes from a professor of that name, so it is spelt with a capital M and abbreviated to just M. The number is the ratio of the speed of an aircraft to the speed of sound under the same conditions. Sound, at sea level, travelling at 762 mph, is taken as Mach 1. So that an aircraft travelling low at 740 mph, less than the speed of sound (subsonic), will be travelling at $740 \div 762$ mph = Mach 0.97. An aircraft travelling at 1,600 mph is at Mach 2.1 ($1{,}600 \div 762$). Since the speed of sound varies with the density and temperature of the air, and so varies according to height, these calculations are not rigid, but near enough for explaining the basis of M numbers.

Another factor which comes into figures quoted for speed, whether in mph, km/h, knots, or expressed as a Mach number, is that of the condition of the aircraft—light or loaded? If external auxiliary fuel tanks are carried the aircraft will be significantly slowed not only by their weight but by their wind resistance and drag—one has only to use an umbrella against a headwind to appreciate the point.

So the speed of an aircraft is rather meaningless unless the condition of the aircraft is known. There is often a wide difference, particularly in the past, between figures quoted by a manufacturer and an operator.

Stalling speed is an important figure for the pilot, the speed at which the aircraft will not be supported in the air. This figure will vary according to loading so, for safety reasons, it is normally given for loaded weight. The figure will vary according whether the flaps are up or down.

Landing speed is related to stalling speed. For safety reasons a low landing speed is desirable, but this must be at a good safety margin above stalling speed. Landing speed is more directly related to landing run, or ground roll as the Americans call it, as the faster the touchdown, the greater the run before coming to a halt. It is here that yards come into use. While reference books such as Jane's keep to feet for measuring take-off and landing runs, many firm's brochures give these figures in yards. This is simply because take-off or landing in 500 yards sounds better than in 1,500 feet.

Range is a figure which can vary by thousands of miles for the same aircraft, all depending on its condition—long or short haul, or being ferried long-distance with its own fuel as its main load. Range may be expressed in miles (not normally abbreviated), kilometres (km to be used instead of miles in the future) or nautical miles (nm—referred to earlier under speed). For a military aircraft a more practical distance of 'radius of action' is quoted. This is the maximum distance from its base in any direction which an aircraft could fly to with its full load, loiter to obtain accurate weapon delivery and, without landing, return to its base.

Endurance is a measure of the time during which an aircraft can remain airborne, expressed in hours and minutes. This is usually to a round figure as seconds do not normally come into it, as it is mandatory to keep a good safety margin of fuel.

Rate of climb in feet per minute and the time it took to reach predetermined heights was among the basic data essential for military aircraft of the past and official trials reports always included this capability.

Ceiling is the height at which rate of climb falls off below 100 feet per minute. This point is sometimes called the service ceiling to differentiate between absolute ceiling—which means just that, the greatest altitude which could be attained.

Specifications for aircraft will include a mass of facts and figures, fin and rudder areas, wheelbase and wheel track, propeller diameter and ground clearance, cabin sizes, baggage space, etc.

Metric/Imperial conversions

Lengths

Inches (in)	×*	25.4 to get millimetres (mm)
Millimetres (mm)	×	0.039 to get inches (in)
Feet (ft)	×	0.0305 to get metres (m)
Metres (m)	×	3.28 to get feet (ft)
Miles	×	1.609 to get kilometres (km)
Kilometres (km)	×	0.621 to get miles
Nautical miles (nm)	×	1.853 to get kilometres (km)
Kilometres (km)	×	0.539 to get nautical miles (nm)
Miles	×	0.869 to get nautical miles (nm)

Areas

Square feet (ft^2)	×	0.093 to get square metres (m^2)
Square metres (m^2)	×	10.764 to get square feet (ft^2)

Velocities

Miles per hour (mph)	×	1.609 to get kilometres per hour (km/h)
Miles per hour (mph)	×	0.869 to get knots (kn)
Knots (kn)		no conversion needed as 1 kn = 1 nautical mile per hour (1 nm/h)
Knots (kn)	×	1.151 to get miles per hour (mph)
Knots (kn)	×	1.853 to get kilometres per hour (km/h)
Kilometres per hour (km/h)	×	0.621 to get miles per hour (mph)
Kilometres per hour (km/h)	×	0.539 to get knots (kn)

Weights

Pounds (lb)	×	0.454 to get kilograms (kg)
Kilograms (kg)	×	2.205 to get pounds (lb)
Short tons	×	0.892 to get long tons
Long tons	×	1.12 to get short tons
Long tons	×	1.016 to get tonnes (t)

Volumes

UK gallons	×	4.546 to get litres (l)
US gallons	×	3.785 to get litres (l)
Litres	×	0.264 to get US gallons
Litres	×	0.22 to get UK gallons
UK gallons	×	1.201 to get US gallons
US gallons	×	0.833 to get UK gallons

*Multiply the figure you want converted by the conversion factor given.

Imperial/metric ready reckoner for quick conversions

Imperial	Metric
1 in	= 25.4 m
1 ft	= 0.3 m
1 mile	= 1.6 km
1 ft^2	= 0.93 m^2
1 gall	= 4.5 l
1 lb	= 0.45 kg
1 mph	= 1.6 km/h
1 mm	= 0.04 in
1 m	= 3.3 ft
1 km	= 0.62 mile
1 m^2	= 10.76 ft^2
1 l	= 0.22 gall
1 kg	= 2.2 lb
1 km/h	= 0.62 mph

Metric prefixes

Mega	= 1,000,000 times or 10^6 (M)
Kilo	= 1,000 times or 10^3 (k)
Hecto	= 100 times or 10^2 (h)
Deca	= 10 times (da)
Deci	= $\frac{1}{10}$ of or 10^{-1} (d)
Centi	= $\frac{1}{100}$ of or 10^{-2} (c)
Milli	= $\frac{1}{1,000}$ of or 10^{-3} (m)
Micro	= $\frac{1}{1,000,000}$ of or 10^{-6} (μ)

Chapter 4

World aircraft industries

This chapter sets the scene of aircraft manufacturing plants over the years and briefly describes the major organisations of today. To show how many famous names of the past, like Avro, Blériot, Junkers, etc, have disappeared into other organisations, charts showing their lifelines and the mergers of national industries have been prepared. There has been a tendency in recent years for large-scale mergers within national industries, and linkage with international consortiums.

International consortiums

Concorde, the world's only supersonic transport aircraft, resulted from collaboration between Britain and France starting some 20 years ago, involving Sud-Aviation and the Bristol Aeroplane Company which are now both known by other names, as the charts show. Some of the world's most important military aircraft are similarly produced by international consortiums. One such is the Tornado, built by Panavia Aircraft GmbH, in one of the largest European aerospace activities yet undertaken, with British Aerospace, Messerschmitt-Bolkow-Blohm (MBB) from Germany and Aeritalia of Italy producing a Multi-Role Combat Aircraft (MRCA) for their respective air arms. The Tornado production programme is controlled by NAMMA (NATO MRCA Management Agency). Another important military aircraft, the Jaguar, is produced by the Anglo-French firm SEPECAT (Société Européenne de Production de l'Avion ECAT) formed by British Aerospace and Dassault/Breguet. MBB, mentioned in connection with the Tornado, also had a connection with VFW-Fokker in Germany and Aérospatiale of France to form Transall, producing the C160 military transport. France and Germany also collaborate to produce their Alpha Jet trainer and a British, French and German consortium form Airbus Industrie to produce the wide-bodied Airbus.

Country briefs

In **America** the aircraft industry has been afflicted with fewer political shackles than in Britain and pre-war names such as Boeing and Lockheed still have a place in today's giant corporations. On the other hand their world marketing is subject to governmental control. Some of the companies have diverse interests of which aircraft manufacture is but one. A complication is the number of aircraft of which designs have been sold to other firms, for example, the Navion designed by North America, who built over a thousand, was sold to Ryan who built more and then sold out to Navion Aircraft. In general the positions on the

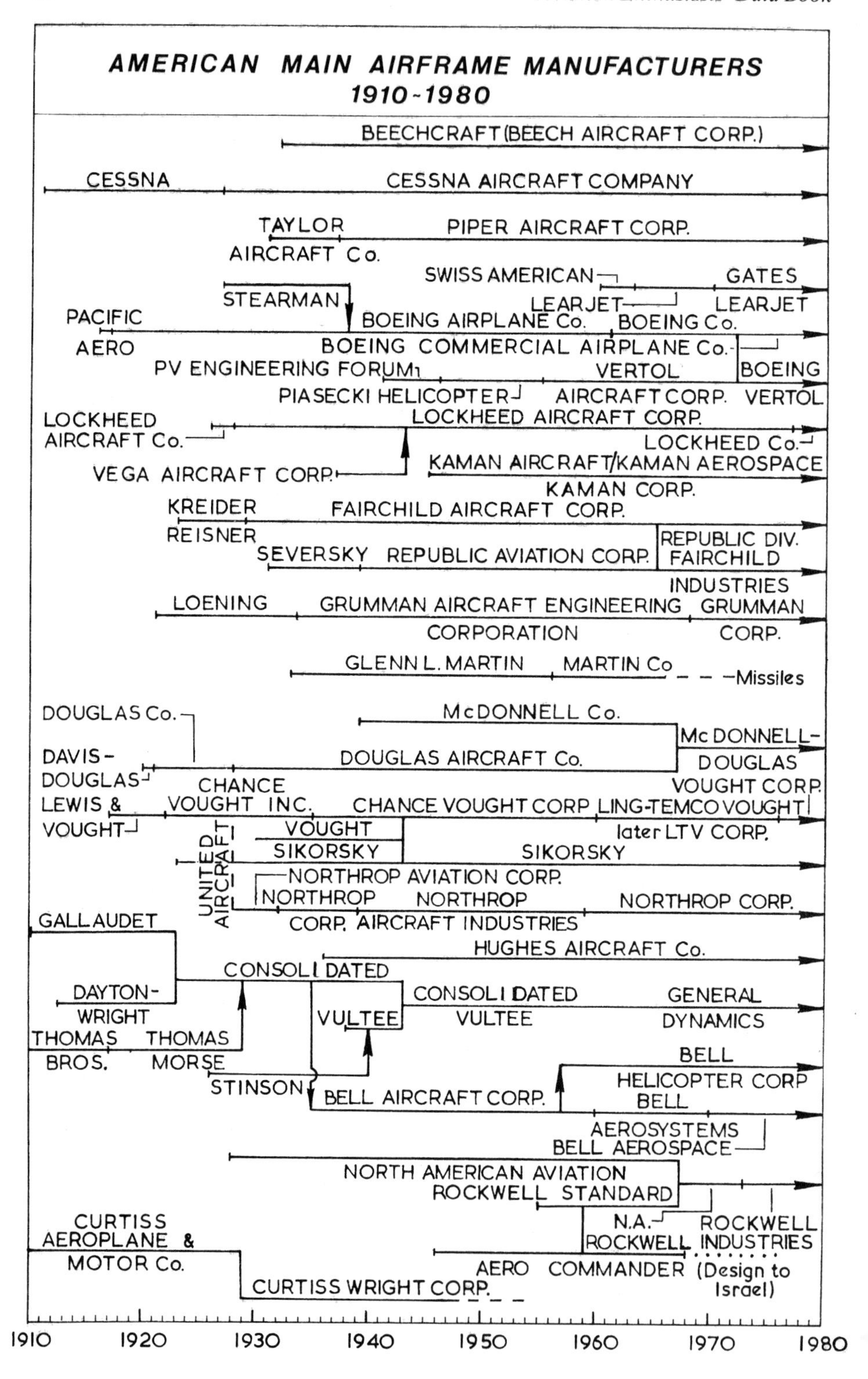
AMERICAN MAIN AIRFRAME MANUFACTURERS
1910-1980
BEECHCRAFT (BEECH AIRCRAFT CORP.)
CESSNA
CESSNA AIRCRAFT COMPANY
TAYLOR
AIRCRAFT Co.
PIPER AIRCRAFT CORP.
SWISS AMERICAN
GATES
LEARJET
LEARJET
STEARMAN
PACIFIC
AERO
BOEING AIRPLANE Co.
BOEING Co.
BOEING COMMERCIAL AIRPLANE Co.
PV ENGINEERING FORUM
VERTOL
AIRCRAFT CORP.
BOEING
VERTOL
PIASECKI HELICOPTER
LOCKHEED
AIRCRAFT Co.
LOCKHEED AIRCRAFT CORP.
LOCKHEED Co.
VEGA AIRCRAFT CORP.
KAMAN AIRCRAFT/KAMAN AEROSPACE
KAMAN CORP.
KREIDER
REISNER
FAIRCHILD AIRCRAFT CORP.
SEVERSKY
REPUBLIC AVIATION CORP.
REPUBLIC DIV.
FAIRCHILD
INDUSTRIES
LOENING
GRUMMAN AIRCRAFT ENGINEERING
CORPORATION
GRUMMAN
CORP.
GLENN L. MARTIN
MARTIN Co
Missiles
DOUGLAS Co.
DAVIS-
DOUGLAS
McDONNELL Co.
DOUGLAS AIRCRAFT Co.
McDONNELL-
DOUGLAS
LEWIS &
VOUGHT
CHANCE
VOUGHT INC.
CHANCE VOUGHT CORP
LING-TEMCO VOUGHT
VOUGHT CORP.
later LTV CORP.
UNITED
AIRCRAFT
VOUGHT
SIKORSKY
SIKORSKY
NORTHROP AVIATION CORP.
NORTHROP
CORP.
NORTHROP
AIRCRAFT INDUSTRIES
NORTHROP CORP.
GALLAUDET
HUGHES AIRCRAFT Co.
CONSOLIDATED
DAYTON-
WRIGHT
THOMAS
BROS.
THOMAS
MORSE
VULTEE
CONSOLIDATED
VULTEE
GENERAL
DYNAMICS
STINSON
BELL AIRCRAFT CORP.
BELL
HELICOPTER CORP
BELL
AEROSYSTEMS
BELL AEROSPACE
NORTH AMERICAN AVIATION
ROCKWELL STANDARD
N.A.
ROCKWELL
ROCKWELL
INDUSTRIES
CURTISS
AEROPLANE &
MOTOR Co.
CURTISS WRIGHT CORP.
AERO COMMANDER
(Design to
Israel)
1910
1920
1930
1940
1950
1960
1970
1980

time scale of the chart reflect the changes of name, rather than the changes of financial controls which may have taken place earlier.

Australia's aircraft industry dates from the years immediately before the Second World War when, with the prospect of a European war affecting defence supplies, a home industry was organised. Aircraft such as the North American Harvard and Bristol Beaufort were put into large-scale production followed by other types. Now the main Commonwealth Aircraft Corporation (CAC) mainly does overhaul work for the Australian Forces which use American, British, French, Italian and Swiss aircraft. The main aircraft production organisation is the Government Aircraft Factories, employing some 2,500 people, currently engaged on building their Nomad twin-turboprop light utility transport.

In **Belgium**, Bollekens of Antwerp were building aircraft before the First World War, but subsequent occupation by Germany in 1914-18 and 1939-44 seriously disrupted their industry. SABCA (Société Anonyme Belge de Constructions Aéronautiques), however, their largest manufacturer since its inception in 1920, still thrives on being involved in large-scale component manufacture as well as assembling Alpha Jets. Avions-Fairey, a branch of Fairey Aviation since 1931, went into liquidation in 1977 and its production of Islanders (as with Britten-Norman in Britain) was taken over by Pilatus of Switzerland. However, the Fairey Gosselies plant has been taken over by SONACA (Société Nationale de Construction Aéronautique) formed with help from the Belgian Government, SABCA and other organisations, and it now participates in the F-16 production programme. Earlier constructors like Renard have disappeared, as have Stampe, with which it joined. But while one would be hard put to find a Renard, even in a museum, Stampe-et-Vertongen SV4 biplanes are still flying and are as much sought after as their British contemporary, the DH Tiger Moth.

In South America both **Argentina** and **Brazil** have aircraft industries and the latter, with their Bandierante light aircraft, have even broken into the North American and European market, including Britain.

Like car manufacture, the **British** aircraft industry has, over the years, been subject to firms amalgamating both for commercial and political reasons. A large amount of work is sub-contracted and this was particularly so during the World Wars. For example, in the First World War the Sopwith Camel was built in thousands by eight sub-contractors, while Sopwith Aviation only built it in hundreds. Between the wars, it was Air Ministry policy to disperse contracts, such as for Hawker Harts, to various firms to keep them in business for defence reasons. During the Second World War there were government-sponsored 'Shadow factories' for both increased production and the dispersal of sites as a precaution against bombing completely disrupting production.

The chart shows how famous names of British industry such as Hawker, Vickers, etc, continue under new organisations, while names like Handley Page came to a dead stop.

There have been three main influences on **Canada**'s aircraft industry which can logically be called A, B and C; America by its proximity, Britain by Commonwealth affiliations, and Canadian enterprise. This is reflected in the pattern of their airframe industry today. There are two main firms, Canadair, which was originally a subsidiary of General Dynamics of America, and de Havilland of Canada (DHC). While DHC was derived from the British de

Havilland parent in 1928, and also became part of Hawker Siddeley (as shown on the British chart), it is now owned by the Canadian Government. However, Hawker Siddeley still flourish in Canada; being known as A.V. Roe (Avro) Canada until 1962, they remain in the Dominion in the aero engine field. Other smaller Canadian companies currently specialise in conversions and amphibians.

The Peoples Republic of China has several aircraft factories, all state owned, which have been building post-1952 to mainly Russian aircraft designs, but they are now having some success with designs of their own.

Czechoslovakia, another Communist country, also has its factories all under state control. The largest, Aero, is as old as the country, but many of the other old manufacturers—Avia, Benes-Mraz, Letov, Praga and Tatra—have disappeared, at least as far as aeronautical work is concerned. In their place is Let, established in 1950, producing transport aircraft. However, the famous Bata shoe company's aeronautical offshoot of Zlin, founded in 1935, still produces aerobatic and touring aircraft in the country.

France once led the world and still has a very important place in world aviation. The decision to nationalise the major part of the aircraft industry into regional units in 1936, splitting some organisations with different locations, is reflected in the chart of the French industry. This reorganisation disrupted their industry at a critical time and, under the German occupation of 1940-4, there was further reorganisation, followed by a post-war reorganisation but still on a regional basis. The present Aérospatiale, achieved by merging Nord and Sud (North and South), was at French Government instigation. The abbreviations shown in the chart of their industry are as follows:

ANF	Aletiers de Constructions du Nord de la France
CAMS	Chantiers Aéro-Maritimes de la Seine
LN	Loire-Nieuport
SEA	Société d'Etudes Aéronautiques
SECM	Société d'Emboutissage et de Constructions Méchaniques
SFCA	Société Française de Constructions Aéronautiques
SFECMAS	Société Française d'Etudes et de Constructions de Material Aéronautiques Speciaux
SIMB	Société Industrielle des Melaux et du Bois
SNCAC	— du Centre
SNCAM	Société Nationale — du Midi
SNCAN	de Constructions — du Nord (the North)
SNCAO	Aéronautiques — de l'Ouest (of the West)
SNCASE	— de Sud-Est (of South-East)
SNCASO	— de Sud Ouest (of South-West)
SPAD	Société pour Aviation et ses Dérives
SPCA	Société Provençale de Constructions Aéronautiques

There have been two large breaks in the **German** aeronautical industry. The Armistice conditions after the First World War not only wrecked the German economy in general, but imposed embargoes on aircraft construction leading to clandestine activities. The chart of the German aircraft industry reflects the scant activity during 1920-30 after the peak of activity in 1915-18, and the increase of tempo when the Nazis came to power in the early '30s leading to another peak period in 1936-45. After that Germany divided into East and West

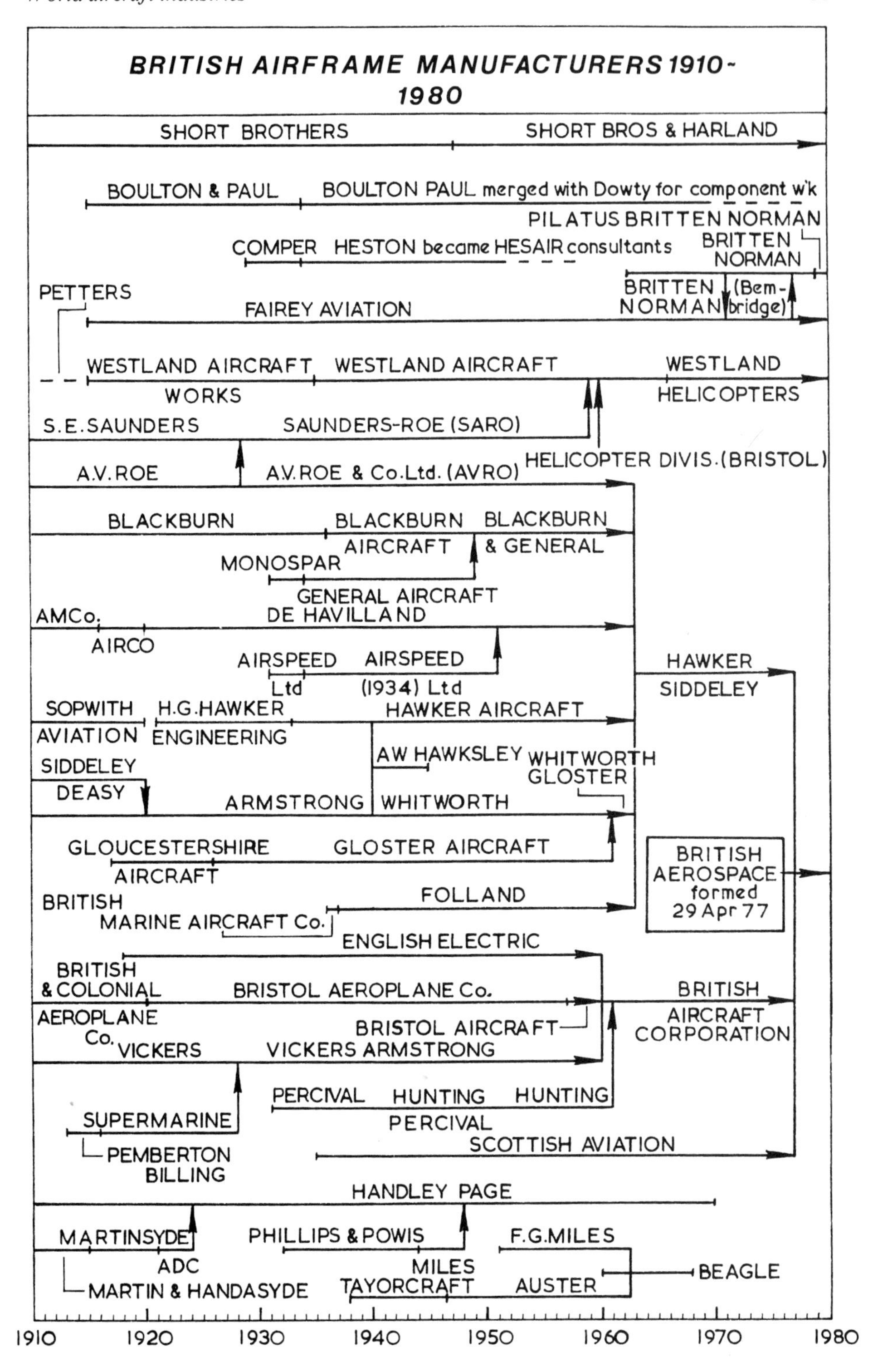
BRITISH AIRFRAME MANUFACTURERS 1910-1980
SHORT BROTHERS
SHORT BROS & HARLAND
BOULTON & PAUL
BOULTON PAUL merged with Dowty for component w'k
PILATUS BRITTEN NORMAN
COMPER
HESTON became HESAIR consultants
BRITTEN NORMAN
PETTERS
FAIREY AVIATION
BRITTEN NORMAN
(Bem-bridge)
WESTLAND AIRCRAFT WORKS
WESTLAND AIRCRAFT
WESTLAND HELICOPTERS
S.E.SAUNDERS
SAUNDERS-ROE (SARO)
HELICOPTER DIVIS.(BRISTOL)
A.V.ROE
A.V.ROE & Co.Ltd. (AVRO)
BLACKBURN
BLACKBURN AIRCRAFT
BLACKBURN & GENERAL
MONOSPAR
GENERAL AIRCRAFT
AMCo.
DE HAVILLAND
AIRCO
AIRSPEED Ltd
AIRSPEED (1934) Ltd
HAWKER SIDDELEY
SOPWITH AVIATION
H.G.HAWKER ENGINEERING
HAWKER AIRCRAFT
SIDDELEY DEASY
AW HAWKSLEY
WHITWORTH GLOSTER
ARMSTRONG
WHITWORTH
GLOUCESTERSHIRE AIRCRAFT
GLOSTER AIRCRAFT
BRITISH AEROSPACE formed 29 Apr 77
BRITISH MARINE AIRCRAFT Co.
FOLLAND
ENGLISH ELECTRIC
BRITISH & COLONIAL AEROPLANE Co.
BRISTOL AEROPLANE Co.
BRITISH AIRCRAFT CORPORATION
BRISTOL AIRCRAFT
VICKERS
VICKERS ARMSTRONG
PERCIVAL
HUNTING PERCIVAL
HUNTING
SUPERMARINE
PEMBERTON BILLING
SCOTTISH AVIATION
HANDLEY PAGE
MARTINSYDE
PHILLIPS & POWIS
F.G.MILES
ADC
MILES
BEAGLE
MARTIN & HANDASYDE
TAYORCRAFT
AUSTER
1910
1920
1930
1940
1950
1960
1970
1980

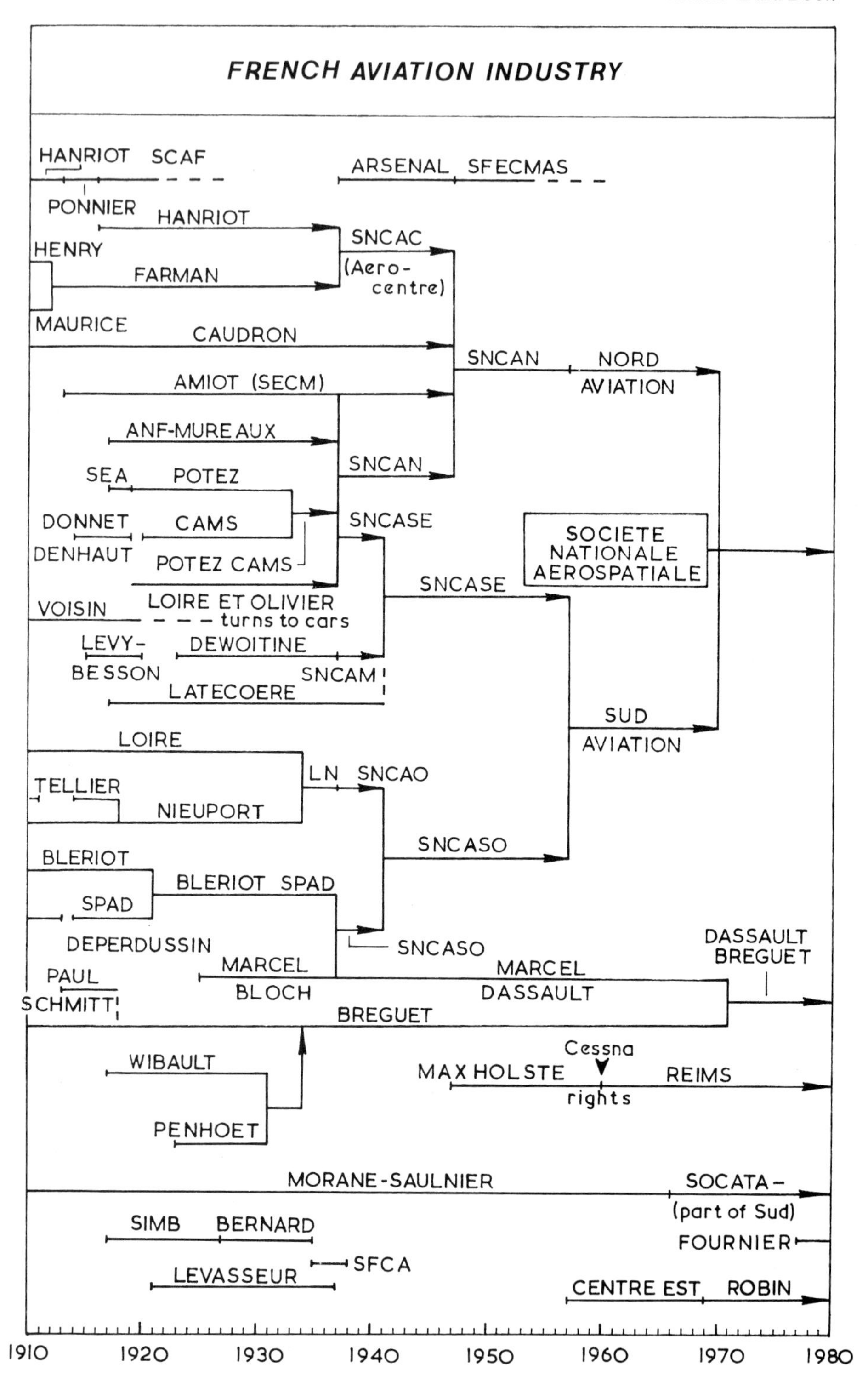
FRENCH AVIATION INDUSTRY
HANRIOT SCAF
PONNIER
HANRIOT
ARSENAL SFECMAS
SNCAC
(Aero-
centre)
HENRY
FARMAN
MAURICE
CAUDRON
SNCAN
NORD
AVIATION
AMIOT (SECM)
ANF-MUREAUX
SNCAN
SEA
POTEZ
DONNET
DENHAUT
CAMS
POTEZ CAMS
SNCASE
SOCIETE
NATIONALE
AEROSPATIALE
SNCASE
VOISIN
LOIRE ET OLIVIER
turns to cars
LEVY-
BESSON
DEWOITINE
SNCAM
LATECOERE
SUD
AVIATION
LOIRE
TELLIER
NIEUPORT
LN
SNCAO
SNCASO
BLERIOT
SPAD
BLERIOT SPAD
SNCASO
DEPERDUSSIN
PAUL
SCHMITT
MARCEL
BLOCH
BREGUET
MARCEL
DASSAULT
DASSAULT
BREGUET
WIBAULT
PENHOET
Cessna
MAX HOLSTE
rights
REIMS
MORANE-SAULNIER
SOCATA -
(part of Sud)
FOURNIER
SIMB
BERNARD
LEVASSEUR
SFCA
CENTRE EST
ROBIN
1910
1920
1930
1940
1950
1960
1970
1980

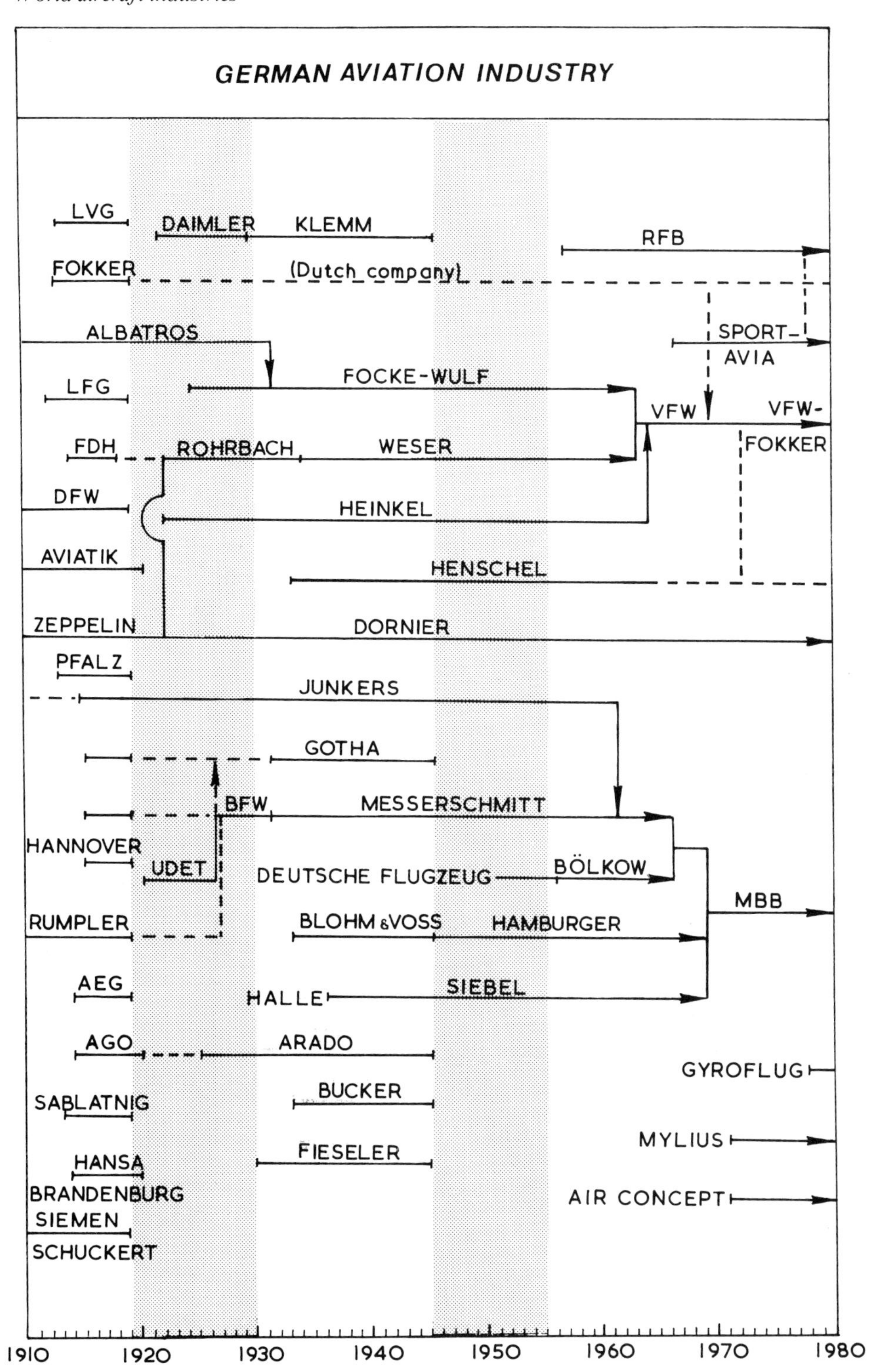

GERMAN AVIATION INDUSTRY
LVG
DAIMLER
KLEMM
RFB
FOKKER
(Dutch company)
ALBATROS
SPORT-
AVIA
LFG
FOCKE-WULF
VFW
VFW-
FOKKER
FDH
ROHRBACH
WESER
DFW
HEINKEL
AVIATIK
HENSCHEL
ZEPPELIN
DORNIER
PFALZ
JUNKERS
GOTHA
BFW
MESSERSCHMITT
HANNOVER
UDET
DEUTSCHE FLUGZEUG
BÖLKOW
MBB
RUMPLER
BLOHM & VOSS
HAMBURGER
AEG
HALLE
SIEBEL
AGO
ARADO
GYROFLUG
SABLATNIG
BUCKER
MYLIUS
HANSA
BRANDENBURG
FIESELER
AIR CONCEPT
SIEMEN
SCHUCKERT
1910
1920
1930
1940
1950
1960
1970
1980

and, under the Potsdam Treaty, was denied military aviation. However, under the 1955 Paris Treaty, the Federal German Republic (West Germany) was brought into the NATO framework of western defence. A new aircraft industry, as the continuity of lines show on the chart, was based on the former organisations but in a way different to any of the other of the world's industries. Because of the devastation during the war, completely new plant had to be built, giving Germany the most modern aircraft factories in the world. Thus, Germany is not merely an equal partner in some large international aeronautical consortiums, but is also the hub of others. Abbreviations used in the German chart are as follows:

AEG	Allgemeine Elektrizitäts Gesellschaft (General Electrical Company)
BFW	Bayerische Flugzeugwerke (Bavarian Aircraft Works)
DFW	Deutsche Flugzeugwerke (German Aircraft Works)
Fdh	Friedrichschafen
LFG	Luftfahrzeug Gesellschaft (Aeronautical Company)
LVG	Luft-Verkehrs Gesellschaft (Air Traffic Company)
MBB	Messerschmitt-Bolkow-Blohm
RFB	Rhein-Flugzeugbau (Rhine Aircraft Factory)
VFW	Vereinigte Flugtechnische Werke (Union of Aircraft Technical Works)

Like other European countries, **Italy** has amalgamated its aeronautical industry into a few large organisations. Over the years it has competed with other countries of the West and, in the past, together with America, was our chief rival in the Schneider Trophy contest. The Italian industry has also been well balanced and was a serious rival in the field of aero engines—as one would expect from a country with an automobile industry of long standing.

Currently most Italian firms have ties with American aviation. Aeritalia, the largest aircraft manufacturer, which took over Fiat (Fabbrica Italiana di Automobili Torino) is to produce 15 per cent of the new Boeing 767 transport. Aermacchi (the Macchi firm founded in 1912) have ties with Lockheed and are to link with Aeritalia in new projects. Agusta, with an aeronautical association going back to 1907, has licence for Bell and Sikorsky helicopter manufacture and Meridionali, formed in 1967 with help from Agusta, has licence rights for Boeing Vertol Chinook helicopters which are being built by SIAI-Marchetti (once known as Savoia-Marchetti and now part of the Agusta group). Apart from the large groups there are several independent firms and among these SSVV (Sezione Experimentale Volo a Vela) are currently building a war-time aircraft in quantity! The Stinson L-5 Sentinel liaison aircraft of the Second World War became popular as a club touring and glider tug aircraft and in the '60s, of some 80 Sentinels registered in Europe, 75 were in Italy. With a continuing demand, SSVV are engaged on conversions to a Super Stinson standard to keep the aircraft in use.

Up to the '30s **Japan** largely followed American and European design trends with help from those two continents, but subsequently their large heavy-engineering firms developed aircraft departments which were quickly expanded. For a period after the Second World War Japan was denied armed forces but, since regaining sovereignty in 1952, Self Defence Forces (including an Air Force) have been formed, initially using American equipment, and maintained by home-produced aircraft mainly built to American designs. The post-war industry has been based on the firms of the Second World War. Fuji Heavy

Industries, building light aircraft and Bell helicopters, is the successor of Nakajima which built nearly 30,000 aircraft up to 1946. Kawasaki and Mitsubishi, which were active during the war, now build a military transport and a close support fighter to their respective designs. Kawanashi, builder of flying-boats in war-time, changed its name to Shin Meiwa in 1949 and has built the only maritime patrol flying-boats currently in service.

Holland's aircraft industry has been represented solely by Fokker-VFW, formed in 1969 by the Royal Netherlands Aircraft Factories Fokker, of some 50 years' standing, and VFW of Germany, but they separated in 1980. Pre-war, the Dutch firms, apart from Fokker, had been Aviolander, which built Curtiss and Dornier aircraft under licence, De Schelde, which in 1934 took over the staff of the former Pander Company, and Frederick Koolhoven who started his own firm in 1934. Koolhoven had designed aircraft since 1910, and he gave his initials FK to Armstrong Whitworth (AW) and British Aerial Transport (BAT) designs of the First World War.

Pre-war PZL (Panstwowe Zaklady Lotnieze—National Aircraft Establishment) was the largest aircraft manufacturer in **Poland**, so that in post-war Communist Poland it was deemed the appropriate authority to control production in all plants in the country producing Polish- and Russian-designed aircraft. The name is now modified to Polskie Zaklady Lotnicze (Polish Aviation Organisation) for which the initials PZL still appertain.

Romania has had an aircraft industry of long standing with the IAR (Industria Aeronautical Romania) works at Brasov being the main plant. Today IAR represents the national industry so its initials are still used as a prefix to the state design organisations which include: ICA-Brasov (Intreprinderea de Constructu Aeronautice—Aircraft Construction Factory at Brasov); IRMA—Bucuresti (Intreprinderea de Reparat Material Aeronautic—Aircraft Component Repair Factory of Bucharest) build Britten-Norman Islanders; and GAB (Grupul Aeronautic Bucuresti—Bucharest Aircraft Group) build BAC 111s, a tribute indeed from a Soviet satellite country. A new joint venture with Yugoslavia is the production of the IAR93/Orao single-seat fighter by CIAR (Centrala Inderstriala Aeronautica Romana—Central Industry for Aircraft of Romania).

After the October 1917 revolution in **Russia**, which incidentally caused Igor Sikorsky, the great helicopter designer then building fixed-wing aircraft, to go to America, all private enterprise was banned. From then on aircraft were built in state factories each known by number.

In **Spain** CASA (Construcciónes Aeronauticas SA—Constructors of Aircraft Ltd), inaugurated in 1923 for building aircraft under licence for the Spanish Forces, now build to their own designs and their Aviocar light utility aircraft is currently gaining export orders.

Saab-Scania, originally the Svenska Aeroplan AB (SAAB) formed in 1937, has become **Sweden**'s prime manufacturer of military and civil aircraft. In **Switzerland** the Federal Aircraft Factory has done much experimental work and erected aircraft for the Swiss Air Force since the '20s. There are now two main Swiss commercial firms, Pilatus, formed in 1939, and FFA (Flug und Fahrzeugwerke AG), formerly the Swiss Dornier works, established in 1926.

Yugoslavia had two main aircraft firms pre-war, Ikarus and Rogojarsky. Now their main firm, owned by the state, is Soko. With Romania, Yugoslavia is collaborating to produce the Orao single-seat tactical fighter, powered by two

Rolls-Royce Viper engines, with production spread around some 30 Yugoslav engineering works. The project was originally known as Jurom (short for Jugoslavia and Romania).

In the Middle East both **Egypt** and **Israel** have relatively new industries. The manufacture of aircraft in Egypt started from co-operation with a German team of engineers in 1950 resulting in the production of several hundred Bücker Bü181 Bestmann primary trainers followed by production of Hispano Ha-200 Saeta jet trainers. The industry was centred on Helwan near Cairo in the '60s to produce a jet fighter. This did not reach the production stage but the facilities at Helwan now provide the base for ABHC (Arab British Helicopter Company) sponsored by the Arab Organisation for Industrialisation and funded jointly by Egypt, Qatar, Saudi Arabia and the United Arab Emirates to produce Westland helicopters. Israel similarly has one main manufacturing organisation, the IAI (Israel Aircraft Industries) which started in 1953 as Bedek Aviation and changed its name to IAI in 1967. Apart from overhaul and component manufacture, IAI produce three main types of aircraft, the Kfir (Lion Cub) jet fighter based on the French Mirage 5, the Arava twin-turboprop twin-boom military transport and the Westwind rear-jet business transport which has a maritime surveillance variant.

Chapter 5

Aircraft national nomenclature

Nomenclature is the system of naming and, with aircraft, this has varied over the years. In the early days of aviation, aeroplanes were classed by their makers' names and a brief description, for example, the Sopwith biplane. When Sopwith had different types of biplane, then 'Sopwith single-seat biplane' would be both the type name as well as the description of one particular type.

Another way was to classify aircraft by the horsepower of their engines, eg, Blériot 50 hp monoplane, to distinguish it from an 80 hp model. At the same time, the aircraft manufacturers were allotting their own type numbers. A.V. Roe (Avro) after using letters to describe his aircraft types, started numbering types from 500 leading to the famous Avro 504 series of the First World War. Sub-types with the same basic airframe but having modified fins or engine changes, were designated 504A, 504B, 504C, et seq. Blériot, in contrast, used Roman numerals to identify types. Other manufacturers used letters or numbers, some prefixed with the designer's initials.

Aircraft production was very limited until the First World War when vast orders were placed. Most of the belligerents introduced their own nomenclature which largely conditioned the national aircraft type naming systems of today.

Britain

By far the most widely used aircraft types in the early years of the First World War were the designs of the Royal Aircraft Factory (which became the Royal Aircraft Establishment after April 1 1918 when the RAF was formed). The British public were prepared for governmental experimental units, but not official production units which was the province of private industry. For this reason the products of the factory had to be called 'experimental' and so aircraft designations like BE2 and FE2, meaning Blériot Experimental for an aircraft nothing like a Blériot and Farman Experimental for an aircraft well removed from a Farman design, came into use. However, only relatively small batches were actually produced at the Farnborough factory; the majority were sub-contracted out to industry.

The aircraft from industry were given type names by the firm which made them, and the Services either adopted this name or used one of their own, eg, in 1916 the design called the Sopwith Pup by Sopwith Aviation, the makers, was officially the Sopwith Single-Seat Scout in the RFC. The Air Department of the Admiralty had the peculiar system of type numbers, based on the individual aircraft identity number of the prototype. Perhaps the best example of this is the

Short 184 floatplane, the type number being derived from No 184, the number of the 184th aircraft and first of this type registered by the Navy. The Sopwith biplane, colloquially called the 1½-Strutter, was Types 9400 and 9700 for two-seat fighter and single-seat bomber versions respectively in naval service, from the serial numbers of the first taken into service.

In late 1917 a Standard Nomenclature was adopted by the Ministry of Munitions and new aircraft from that time were named according to its guidelines. Fighters were given, at first, names of animals, whilst bombers received place names. In general the names chosen were alliterative with the firm's name, eg, Sopwith Snipe, Vickers Vimy, Westland Wagtail. Where there were variations of a basic type, such as an engine change, mark numbers in Roman numerals were introduced. The first major change came in February 1942 when significant role letters, such as B for Bomber and F for Fighter were prefixed to the mark number; thus Spitfire Mk V would become officially Spitfire F Mk V, usually expressed as just Spitfire FV.

Role letters which have been allotted over the years, given here in alphabetical order, are: AEW—Airborne Early Warning; AH—Army Helicopter; AL—Army Liaison; AOP—Air Observation Post; AS—Anti-Submarine; B—Bomber; B(I)—Bomber (Interdictor); B(K)—Bomber (Tanker); B(PR)—Bomber (Photo Reconnaissance); C—Cargo or passenger transport; CC—Cargo and Communications; COD—Carrier Onboard Delivery; D—Drone or pilotless (replaced U); DW—mine-exploding (non-significant letters for security reasons); E—Electronics (including flight-checking equipped); F—Fighter; F(AW)—Fighter (All-Weather); FB—Fighter-Bomber; FG and FGA—Fighter Ground Attack; FGR—Fighter Ground-Attack and Reconnaissance; FR—Fighter Reconnaissance; FRS—Fighter Reconnaissance Strike; GA—Ground-Attack; GR—General Reconnaissance (until 1950 when replaced by MR) later Ground-Attack Reconnaissance; HAR—Helicopter Air Rescue; HAS—Helicopter Anti-Submarine; HC—Helicopter Cargo; HF—High (altitude) Fighter; HR—Helicopter Rescue; HT—Helicopter Training; HU—Helicopter Utility; K—Tanker (K chosen to comply with US usage and the fact that T was allotted for Trainer); LF—Low (altitude) Fighter; MR—Maritime Reconnaissance; O—Observation; PR—Photographic Reconnaissance; R—Reconnaissance; S—Strike; SR—Strategic Reconnaissance; T—Trainer; TF—Torpedo Fighter; TT—Target Tower; TX—Training Glider; U—Utility; W—Weather.

Earlier, in 1940, a suffix letter had been introduced to classify armament, eg, a Spitfire FVa had the original eight .303 Browning guns, an FVb had two 20 mm cannon, and those fitted with a universal wing permitting varied fittings became the FVc. Later suffix letters to the mark numbers were also used to identify other than armament modifications.

In 1942, when Roman figures became rather unwieldy for high mark numbers, those over 20 were expressed in standard Arabic figures and in 1947 Arabic figures became standard for all mark numbers of British Service aircraft. The official nomenclature notices issued by the MOD(PE) give the form Hercules C Mk 3, which everyone, except apparently the officials, simply calls the Hercules C3.

British firms have had their own type numbering and naming systems which, owing to amalgamations, have resulted in anomalies. The Avro type number series, of which type 683 was the Lancaster, had the Avro 748 under development when Hawker Siddeley took over, so it became the HS748. Similarly de Havilland's DH series, staring in the First World War and including such milestones as the DH106 Comet, had their DH125 light rearjet transport taken over as the HS125. With Hawker Siddeley now in British Aerospace, and both HS125 and HS748 designs still in production, these aircraft are being referred to as BAe125 and BAe748. These changes can be very confusing unless you get the picture from an industry chart, as given earlier. Details of the types produced by British firms of the past, with their designation systems including projects, with a history of the firms concerned are contained in the excellent series published by Putnam, all in uniform size, covering: Airspeed, Armstrong Whitworth, Avro, Blackburn, Bristol, de Havilland, Fairey, Gloster, Handley Page, Hawker, Miles, Short, Supermarine and Vickers. Ian Allan produced a briefer *Aircraft Album* series in the early '70s, but the range of British firms only included Handley Page and Hawker. Westland was covered by Ian Allan in another series.

Canada

In the past the Canadians have used the designations and names appropriate to the American and British aircraft which they had in service. Of the Canadian-built aircraft, de Havilland of Canada have prefixed their designs DHC in a simple numerical series using, at first, names of Canadian mammals, viz DHC-1 Chipmunk, DHC-2 Beaver, DHC-3 Otter, DHC-4 Caribou, DHC-5 Buffalo, DHC-6 Twin Otter and DHC-7 Dash 7. Canadair Limited, using the appropriate prefix letters CL, allotted, among other type numbers, CL13 for licence-built F-86 Sabres, CL215 for an amphibian and CL600 for their new Challenger.

A Service nomenclature, based on the USAF system, was started by the RCAF shortly after the Second World War. This became the national Service system in 1966 when the RCAF lost its title on integration with other arms and became part of the Canadian Armed Forces (CAF). Role prefix letters were used starting with C for Canadian with a further letter denoting functions as follows: CC—Transport; CF—Fighter; CH—Helicopter; CP—Patrol (Maritime); CSR—Search and Rescue; CT—Trainer. These role letters were hyphenated to service type numbers starting at 100 with the logical exception of the F-18 Hornet coming into service as the CF-18. The Canadians in some cases also bestowed a Service name.

The allocations of Service type numbers to aircraft, with Service names where applicable, have been as follows: CF-100—Avro CF-100; CF-101—McDonnell F-101 Voodoo; CC-102—Avro Jetliner; CF-104—Lockheed F-104 Starfighter; CF-105—Avro Arrow; CC-106—Canadair CL44 Yukon; CP-107—Canadair CL28 Argus; CC-108—DHC-4 Caribou; CC-109—CL-66 Cosmopolitan; CH-112—Hiller Nomad; CH-113—Vertol 107 Labrador; CH-113A—Vertol 107 Voyageur; CT-114—Canadair CL-41 Tutor; CC-115—DHC-5 Buffalo; CF-116—Northrop CF-5; CC-117—Sud Falcon; CH-118—Bell Iroquois; CP-121—Grumman Tracker; CRS-123—DHC-3 Otter; CH-124—Sikorsky S-61 Sea King; CT-129—Douglas DC-3 Dakota; CC-130—Lockheed Hercules; CC-132—DHC-7 Dash 7; CT-133—Lockheed Silver Star; CT-134—Beech Musketeer; CH-135—Bell Twin-Huey; CH-136—Bell Kiowa; CC-137—Boeing 707; CC-138—DHC-6 Twin Otter; CP-140—Lockheed Aurora; CH-147—Boeing Vertol Chinook.

Germany

German firms during the First World War had their own designation systems for the aircraft types they produced and these were used by the German Navy. But the German Army insisted on its own type designating system and in many cases this was marked on the fuselage. The 40 aircraft designing firms during that war were each allotted significant abbreviations, eg, Alb for Albatros Werke, DFW for Deutsche Flugzeugwerke, Fok for Fokker, etc.

Aircraft types were divided into classes designated by a letter allotted alphabetically at first, and later by significant letters in the range: A—unarmed two-seat monoplanes; B—unarmed two-seat biplanes; C—armed two-seat biplanes; CL—close support types; D—single-seat biplanes (but including monoplanes in 1918); Dr—*Dreidekkern* (triplanes); E—single-seat monoplanes; F—Fokker DrI triplane initial production; G—*Grossflugzeug* (Meaning big aircraft; applicable to twin-engined types); J—armoured aircraft for ground support; N—night-bombing C-types; R—*Riesenflugzeug* (giant aircraft).

Each different type, within the classes described above, was designated by Roman numerals. Thus successive types of armed two-seat biplanes by Albatros would be designated Alb CI, CII, CIII, CIV, etc. Minor changes such as different engine in the same airframe were designated by suffix letters, eg, CIa, allotted alphabetically. Many aircraft were subcontracted and the form Halb CLIV(Rol) would refer to the fourth (IV) close support (CL) aircraft type of Halberstadt (Halb) design built by Roland (Rol). Roland was the trade-name of LFG (Luftfahrzeug Gesellschaft) to avoid confusion with LVG (Luft-Verkehrs Gesellschaft). The system lapsed after 1918. One of the most comprehensive books on this subject is *German Aircraft of the First World War* by Peter Gray and Owen Thetford, published by Putnam in 1962 with a second edition in 1970. For a detailed history of the R-types, *The German Giants* by C.W. Haddow and Peter M. Grosz, published by Putnam in 1962 with an enlarged and revised edition in 1969, is recommended.

Germany, under the Third Reich, started a new national system in the '30s for service aircraft. The existing German aircraft firms already had their separate type letter and numbering systems, when the new national system came into general use with type numbers prefixed by two letters signifying the design firm, viz: Ar—Arado; Bf—Bayern Flugzeugwerke (Bavarian Aircraft Works, which was renamed Messerschmitt from July 11 1938); Bu—Bucker; BV—Blohm und Voss; Do—Dornier; Fi—Fieseler; Fl—Flettner; FW—Focke-Wulf; Go—Gotha; He—Heinkel; Hs—Henschel; Ho—Horten; Ju—Junkers; Ka—Kalkart; Me—Messerschmitt; Si—Siebel; Ta—Tank (for designer Kurt Tank). One uses, therefore, Bf108, Bf109 and Bf110, not the 'Me' prefix, since these designs originated before the change of company name. Following in the number sequence the He111 was built in quantity. Numbers like He119 are less familiar because they refer to more obscure aircraft; in this case a high-speed reconnaissance bomber of which only five were built. Of the allocations, He113 was a propaganda fake to mislead British intelligence.

Modifications, the equivalent of the British mark numbering system, were denoted by suffix letters, allotted alphabetically in the form Bf109E. Further modifications in fittings or armament, not warranting another letter reserved for major changes such as a higher-powered engine, were notified by suffixed numbers, from 1 upwards in the form Bf109E-3. For a complete redevelopment of an aircraft type, a new type number was issued based on the old, eg, the greatly improved Ju88 was allotted Ju188 and further designs on the basic configuration were Ju288, Ju388 and Ju488. The system ended in 1945 and has not been replaced by any further national nomenclature.

Among the comprehensive works, within a single cover, on German aircraft types of the Second World War, is *Warplanes of the Third Reich 1933-45* by William Green, published by Macdonald & Jane's and containing 2,000 photos, and *German Aircraft of the Second World War* by J.R. Smith and Antony King, published by Putnam.

Japan

It is not possible within the scope of this book to give a complete run-down on the Japanese type nomenclature which includes designations and names, as well as different systems for Army and Navy, with varying abbreviated designations. An added complication is that Army aircraft type numbers were based on the year of procurement, eg, Army Type 99 was procured in the Japanese year 2599 which was AD 1940 in our reckoning. The systems are set out in detail in *Japanese Aircraft of the Pacific War* by R.J. Francillon, published by Putnam in 1970. However, as this author states, it is by the Allied code names that the West remembers and refers to the Japanese aircraft of the Second World War, and this is explained.

Owing to lack of knowledge of Japanese aircraft in 1942, a code system was adopted in the South-West Pacific and South-East Asian areas. The code was allotted on the basis of: male names for fighters and reconnaissance floatplanes; female names for bombers including dive-bombers and reconnaissance aircraft whether single-, twin- or multi-engined, and flying-boats; female names with the initial letter 'T' for transports; tree names for trainers; and bird names for gliders. Primarily the system was introduced for identification, but it came to be used particularly for intelligence purposes in computing a Japanese Order of Battle. A 'mark number' system was adopted in 1944 using the Japanese type number as a mark number to the Allied code name.

The code letter allocations were controlled by the Technical Air Intelligence Center, Naval Air Station Anacostia, USA, which urged the avoidance of using the popular unofficial name Zero to describe various fighter types. There was inevitably a shortfall in intelligence resulting in names being given to designations of aircraft which did not exist, or aircraft which did exist but were not in the theatre. On the other hand some types flown by the Japanese did not reach the eyes or ears of Allied Intelligence. A listing of these code names is given in the appendices.

United States

The US Army first promulgated a standard nomenclature for aircraft in September 1919, using a letter-number designating system—not a naming system. Aircraft were required for service in 15 categories, type classified by significant letters hyphenated to the type number within that class, starting at 1 in each case. The 15 type categories in numerical order were: PW—Pursuit Water-cooled engine; PN—Pursuit Night; PA—Pursuit Air-cooled engine; PG—Pursuit Ground-attack; TP—Two-place Pursuit; GA—Ground-Attack: IL—Infantry Liaison; NO—Night Observation; AO—Artillery Observation; CO—Corps Observation; DB—Day Bombardment; NBS—Night Bombardment Short-distance; NBL—Night Bombardment Long-distance; TA—Training Air-cooled engine; TW—Training Water-cooled engine. It will be seen that the word 'pursuit' was used for what the Germans called a 'hunter' *(Jagd)* and the British a 'fighter'. Four special categories were later introduced as follows: A—Ambulance; M—Messenger (liaison); R—Racer for competition flying; T—Transport.

Two important facts should be kept in mind with these and subsequent designations (which are still the basis of the American Service designating system today). Firstly, it is an official system and is irrespective of company considerations and, secondly, the hyphen is an important part of the designation. The basic USAAC, USAAF, USAF and all American Service designations since 1962 consist of a letter hyphenated to a number. To enlarge on this, a significant letter (not one allotted alphabetically) is hyphenated to a number in a numerical series denoted by the letter. Other letters may be prefixed, or suffixed, but they will have different meanings according to which side of the hyphen they are placed. The hyphen is the centre point and is as important in the American designating system as a decimal point in a numerical system. When tabling American types they should be lined up by the hyphen, not by the first letter. Unless you have at the core of any American Service designation (excepting US Navy to 1962) a letter hyphenated to a number, then something is wrong with the designation—as you will often find in newspapers.

Suffix letters were allotted for variations to the basic type, corresponding to British mark numbers. For example the first PW-1 had tapered wings; when fitted with Fokker wings it became the PW-1A and with a further wing change it became the PW-1B. Letters are still allotted for changes, with the difference that now the first model automatically becomes the 'A' version, eg, the first F-15 Eagle of today was the F-15A. The first of the prefix letters was introduced with this early series using X for Experimental, eg, XPW-8A. That this was a Service series and not a manufacturer's series is borne out by the fact that it was the Fokker PW-7, Curtiss PW-8, Boeing PW-9; ie, they were the seventh, eighth and ninth Water-cooled Pursuit types of the US Army, not of a manufacturer.

However, a change came in 1924. The system remained the same, but a new series of type numbers started with a different series of significant prefix letters, known as Basic Mission Symbols, as follows: A—Attack (light bombardment); AT—Advanced Trainer; B—Bombardment; BC—Basic Combat; BT—Basic Training; C—Cargo and transport with UC for Utility Transport used (as appropriate) within the same number series; F—Photographic Reconnaissance; FM—Fighter Multiplace; G—Giroplane; HB—Heavy Bombardment (used 1926-27 only); L—Liaison; LB—Light Bombardment (not used after 1930); O—Observation; OA—Amphibian; P—Pursuit (Fighter); PB—Pursuit Biplane (used mid-'30s only); PT—Primary Trainer. Later R was used for Rotary wings (helicopters) and was replaced later still by H for Helicopter. V is now used for V/STOL (Vertical/Short Take-Off and Landing) and X for experimental projects.

In addition to the X for experimental, introduced as a prefix to the first role letters, other prefixes were introduced from 1928 and became known as Modified Mission and Status Symbols. For example, a B-17G with bombardment (the American term for bombing) as its basic mission, signified by the B, when modified as a transport became a CB-17G since C signified Cargo and transport, as given above. The main modified mission letters which are still in use (together with one or two recently dropped) are: A—Attack; C—Transport; D—Drone Controller/Director; E—Electronic installation;

G—Grounded permanently; H—Search and rescue; J—Temporary special test status; K—Tanker; L—Cold weather operating; M—Missile carrier; N—Permanent special test status; O—Observation; P—Passenger transport (not used since the mid-'60s); Q—Drone radio-controlled; R—Restricted used until 1947 and then signified Reconnaissance; S—Search and rescue until early '60s and then anti-submarine; T—Trainer; U—Utility; V—VIP (Very Important Person or transport for administrative authorities); W—Weather; X—Experimental; Y—Service Test; Z—Obsolete. The status symbols of G, J, N, X, Y, Z, would always appear before the other Mission Symols indicating role; eg, if the CB-17G referred to above was used temporarily for special tests, it would become a JCB-17G.

So much for the prefixes; but in addition to the letter suffixes to denote the equivalent of mark numbers there are sub-series numbers given in an exclusive American way, known as 'block numbers' (the British would probably have called them batch numbers). As aircraft leave production lines, so the modifications in the light of operating earlier models, or other improvements to reduce weight or improve efficiency, are incorporated. The standard and equipment of an early production and a late production aircraft would vary considerably even for the same sub-type, eg, a B-17G. So production block numbers were introduced from 1941 starting at 1 and then issued in an arithmetical progression of 5, ie, in the form B-17G-1, B-17G-5, B-17G-10, B-17G-15. Thus, if there was a further change on the production line, the aircraft was recorded and marked as a B-17G-20 and the appropriate B-17G handbooks and documentation would specify the precise standard of a -20 block. The reason for the progression in fives, was to permit intervening numbers to be used to record field modifications.

Another suffix was introduced at this time, to denote the actual plant in which the aircraft was built. There were well over a hundred plants building complete airframes and some aircraft of the same type were being built in more than one plant. A two-letter code was introduced to identify the particular plant as follows: AD—Air Design; AE—Aeronca; AG—Air Glider; AH—American Helicopter; BA—Bell (Buffalo); BF—Bell (Fort Worth); BH—(Beech); BL—Bellanca; BN—Boeing (Renton); BO—Boeing (Seattle); BR—Briegleb; BS—Bowlus; BU—Budd; BV—Boeing-Vertol; BW—Boeing (Witchita); CA—Chase (West Trenton); CC—Canadian Commercial; CE—Cessna; CF—Convair (Forth Worth); CH—Christopher; CK—Curtiss-Wright (Louisville); CL—Culver; CM—Commonwealth; CN—Chase (Willow Run); CO—Convair (San Diego); CR—Cornelius; CS and CU—Curtiss-Wright (St Louis and Buffalo); DA—Doak; DC and DE—Douglas (Chicago and El Segundo); DH—de Havilland of Canada; DJ—SNCASO of France; DK and DL—Douglas (Oklahoma and Long Beach); DM—Dorman; DO and DT—Douglas (Santa Monica and Tulsa); FA and FS—Fairchild (Hagerstown and Burlington); FE—Fleet; FL—Fleetwings; FO—Ford; FT—Fletcher; GA—GLA; GC—General Motors (Cleveland); GE—General Aircraft; GF—Globe; GK and GM—General Motors (Kansas and Detroit); GN—Gibson Refrigerator; GO—Goodyear; GR—Grumman; GT—Grand Central; GY—Gyrodyne; HE—Helio; HI—Higgins; HL—Hillier; HO—Howard; HU—Hughes; IN—Interstate; KA—Kaman; KE—Kellet; KM—Kaiser; LK—Leister-Kauffman; LM and LO—Lockheed (Marietta and Burbank); MA—Glenn Martin (Baltimore); MC—McDonnell (St Louis); MD and

MF—Martin (Baltimore and Orlando); MH—McCulloch Motors; MM—McDonnell (Memphis); MO—Glen Martin (Omaha); NA and NC—North American (Inglewood and Kansas City); ND—Noorduyn; NF, NH and NI—North American (Fresno, Columbus and Downey); NK—Nash-Kelvinator; NO—Northrop; NT—North American (Dallas); NW—North-western; PH—Piasecki; PI—Piper; PL—Platt LePage; PR—Pratt, Read and Co; RA—Republic (Evansville); RD—Read-York; RE—Republic (Farming-dale); RI—Ridgefield; RO—Robertson; RP—Radioplane; RY—Ryan; SE—Seibel; SI—Sikorsky; SL—St Louis; SP—Spartan; SW—Schweizer; TA—Taylorcraft; TG—Texas Engineering (Greenville); TI—Timm; TP—Texas Engineering (Grand Prairie); UH—United Helicopter; UN—Universal Moulded Products; VE—Vega; VI—Canadian Vickers; VL—Vertol; VN—Vultee (Nashville); VO—Chance Vought; VU and VW—Vultee (Downey and Wayne); WA—Ward Furniture; WI—Wichita Engineering; WO—Waco.

The manufacturer's code is used as a suffix, eg, B-17G-25-BO or, if the block number is not quoted, as B-17G-BO.

In 1962 the system being used by the USAF was made universal for all American Service aircraft, bringing the US Navy and Marine Corps and Army Aviation in line. At the same time the series numbers over 100 were started again. The Cargo/Transport series had reached C-142 in allocations, so the series was started again at 1. This explains why the C-141 Starlifter transport was followed by the C-5 Galaxy which received its type designation in the new series. Similarly the F series was restarted after reaching F-111 in 1962, which is why the recent Hornet has the low F-18 allocation. The A for Attack series was also re-started.

The American military aircraft types and a detailed explanation of the designating system, with all its variations, is contained in the book *United States Military Aircraft Since 1908* by Gordon Swanborough and Peter M. Bowers, published by Putnam 1963 and revised and updated in a 1971 edition.

Until 1962 the US Navy had its own system of aircraft type designations which underwent four changes and is too complicated to detail here. In the main the US Navy system 1922-62 involved letters denoting the function of an aircraft, except in the case of helicopters, gliders and experimental types, when the respective letters H, L and X would precede them. The function letters were of one or two letters with the single exception of PTB for Patrol Torpedo Bomber. In general they were significant letters, eg, OS for Observation Scout. The main exception was that R denoted Transport; the reason being that although T was allotted for transports 1927-30, it was also used for Trainer and Torpedo.

The next letter in the sequence was a Manufacturer's Identification Letter; a single letter in general except for foreign (to America) manufacturers which had two letters. However, this changed according to period and the letter 'E' changed nine times over the years. A complication to the system was a number interposed between the function and the manufacturer's letter, if it was the second, third, etc, type with that function produced by the same manufacturer. To give a simple example, the first Douglas Skyraiders were designated AD (Attack aircraft by Douglas—being the function and maker letters). When Douglas produced their Skyshark attack aircraft this became the A2D (2nd Attack type by Douglas) and the following Skywarrior, the A3D.

The first model of any one type was notified in the form -1, so that the first model of the AD would be AD-1 and subsequent main modifications AD-2, AD-3, et seq. Types modified for other than their basic roles had a suffix letter to this number, which varied in meaning according to the type of aircraft. For example, in the designation PBY-5A the A suffix stood for Amphibian version, on the F4F-3A it stood for landplane version of a carrier aircraft and A also had other meanings with different aircraft types. However, some letters had a single meaning and these were: K—drone conversion; M—Missile launcher; N—Night or all-weather; P—Photographic; Q—ECM; R—transport; S—anti-Submarine; T—two-seat Trainer conversion; U—Utility; W—Warning or search radar fitted; Z—special transport.

However, in 1962, the Navy was brought into line with the national Service aircraft system and their aircraft in service at that time were redesignated. For example, the Crusader, previously designated F8U-1, became the F-8A and the F8U-1P became the RF-8A to fit in the system described for the USAF, which became the national Service aircraft system. Similarly, the Army's HU-1 helicopter had the letters reversed to UH-1 to bring it into line.

A comprehensive book on US naval aircraft types with a full and detailed description of the designating system is entitled *United States Navy Aircraft since 1911* by Gordon Swanborough and Peter M. Bowers.

Irrespective of the Service designations, aircraft manufacturers in America have had their own designating and naming systems. Boeing, who have given their aircraft type numbers since the '20s, caught the public imagination with Flying Fortress as the name of the bomber which the USAAC designated the B-17, and Fortress was even adopted as its official name by the RAF who used it mainly on maritime work. Post-war, following the success of their Model 707 airliner, the firm have adopted the unusual 727, 737, 747 and 757 sequence. Douglas aircraft made the first successful flight round the world in 1924, but it was in the '30s that the company really came to fame with their Douglas Commercial (DC) series of which their DC-3 became the C-47 of the USAAF and was named by them Skytrain, but it is by the British Service name for the type, Dakota, that it is better known to us. The DC series continues today in its DC-8, DC-9 and DC-10 forms. Of the last-named its Service tanker version, the Extender, has taken up the appropriate USAF designation KC-10A.

Grumman, who give a G-prefixed type number to its aircraft, used theme names of Wildcat, Hellcat, Bearcat for successive designs. Lockheed adopted a 'Star' theme viz, Shooting Star, Starfire, Starfighter, Galaxy and now TriStar for their three-engined airliner. Republic had a run of 'Thunder', viz, Thunderbolt, Thunderjet, Thunderstreak and Thunderchief. When it comes to individual firms, there is no universal nomenclature.

USSR

Before Soviet aircraft type nomenclature is reviewed, the reader should appreciate that in the Russian language only the letters A E K M O T are common to English and Russian in form and pronunciation, that B E P C Y X are common in form but are pronounced differently, and that Russian uses 17 Cyrillic characters not in our adopted Roman alphabet. For this reason, to put Russian into English there has first to be a transliteration from Cyrillic to Roman characters before translation.

The first state aircraft nomenclature was a simple system of letters indicative of the aircraft's role, hyphenated to a type number in that role, with some exceptions when a designer's name or a design institute was used. These are given alphabetically: ANT—Andrei Nikolaevich Tupolev (designer); B—*Bombardirovshchik* (Bomber); BDP—*Boevoi Desantnyi Planer* (military assault glider); BSh—*Bronirovannyi Shturmovik* (armoured ground-attack); DB—*Dalnii B* (long-range bomber); DDBSh—(meaning twin-engined long-range armoured ground-attack); DI—*Dvukhmestnyi Istrebitel* (two-seat fighter); GM—*Gelicopter Mil* (helicopter designed by Mil); I—*Istrebitel* (fighter); I-Z—*Istrebel Zvemo* (linked aircraft, ie, carried); ITP—*Istrebitel Tyazhelyi Pushechnyi* (fighter cannon-armed); *LaGG*—Lavochkin, Gorbunov and Gudkiv (designers); M—*Morskoi* (Maritime); MDB—*Morskoi Dalnii Razvedchik* (Maritime long-range reconnaissance); MP—*Motoplaner* (motorised glider); NB—*Nochnoi B* (Night Bomber); PB—*Pikiruyushchii B* (dive-bomber); PI—*Pushka I* (cannon-armed fighter); OKB—*Opytno Konstruktorskoe Byuro* (Experimental Design Bureau); R—*Razvedchik* (reconnaissance); SB—*Skorostnoi B* (fast bomber); TA—*Transportnaya Amfibia* (Transport, Amphibious); TB—*Tyazhëlyi B* (heavy bomber); TIS—*Tyazhelyi I Soprovozh-deniya* (heavy fighter for escort); TsKB—*Tsentral'noe Konstruktorskoe Byuro* (Central Design Bureau); U—*Uchehnyi* (trainer); VI—*Vysotnyi I* (high-altitude fighter); VIT—*Vozduschnyi I Tankov* (destroyer fighter of tanks).

In the '40s a new system was introduced using the initial letters of the design bureau, normally named after the chief designer as in Tu for A.N. Tupolev. This was then hyphenated to a design type number, eg, Tu-16. The type numbers are in general allotted in numerical sequence, but there are some anomalies. The initial letters used over the past 40 years with the name which they stand for are: An—Antonov; Be—Beriev; Che—Chetverikov; Il—Ilyushin; Ka—Kamov; La—Lavochkin; Li—Lisunov; MiG—Mikoyan and Gurevich; Mi—Mil; Mya—Myasishchev; Pe—Petlyakov; Po—Polikarpov; Su—Sukhoi; Tu—Tupolev; Yak—Yakolev.

Variants were notified in some cases in the French way by *bis* appearing as a suffix to the type number, but in general suffix letters are widely used. In the case of the Tu-104A and Tu-104B it is a simple case of denoting the sequence of variants to the basic Tu-104, but in some transliterated designations V is given for the Russian B. Suffix letters, apart from those allotted in sequence, are the initial letter of the Russian word indicative of the special function, condition or equipment of the aircraft. The letter may indicate a different word for different aircraft, but in most cases the appropriate word is self-evident.

The main suffix letters to Soviet aircraft type numbers used from the '40s to today are: D—*Dal'nyi* (long-range, ie, extra tankage); F—*Forsazh* (boosted, eg, afterburner); K—*Kran* (Crane) for a lifting version of a helicopter but *Krorotkonogii* (short-legged) in the case of the Mi-10K flying crane; L—*Lesookhranenie* (forestry protection); M—*Modifikatsyi* (Modified); MF—combination M and F; P—various meanings for passenger or fire-fighting versions, or airborne-radar-equipped in the case of fighters; PF—combination of P and F; R—*Reativny* (jet for jet assisted) or *Razvedchik* (Reconnaissance); S—words meaning spray-equipped or variable incidence tailplane; SBS—words meaning boundary layer blowing; T—words meaning heavy, target-towing or torpedo-carrying versions; TK—*Tonkoe Drylo* (thin wing); U—*Uchebno* (trainer) in general but *Usilennyi* (strengthened) in the case of the Yak-9U; UTI—*Uchebno-Trenirovochny Istrebitel* (trainer two-seat fighter); W has been used in connection with float-equipped versions and Z for weather reporting.

Chapter 6

National insignia

Aircraft of the armed forces of every country in the world carry insignia indicating their nationality. The roundel of the Royal Air Force (RAF) will be familiar to all. The French called it a *cockade* when they adopted it in 1912. In the early months of the First World War the Royal Flying Corps (RFC) were forced to paint Union Jacks under the wings of their aircraft which were attracting ground fire from their own side. But this did not prove practicable for, at a distance, colour is subordinate to shape. The central red St George's Cross of the Union Jack was confused with the German black Iron Cross marking adopted by the German Air Service. So from early November 1914 the RFC adopted the French cockade, but with the order of the colours reversed. This was logical since the national colours of both Britain and France were red, white and blue. This was also suitable for the Americans when they took the field in Europe during 1917-18, with a further re-arrangement of these national colours on their aircraft as illustrated further on. The Royal Naval Air Service (RNAS), which had used a red ring in place of the Union Jack from December 1914, adopted the RFC roundel from November 1915 and this then became the standard British insignia.

In model kits and decal sheets, as well as articles on markings, the changes over the years in British roundels have been classified by a letter denoting variations in proportions, as illustrated.

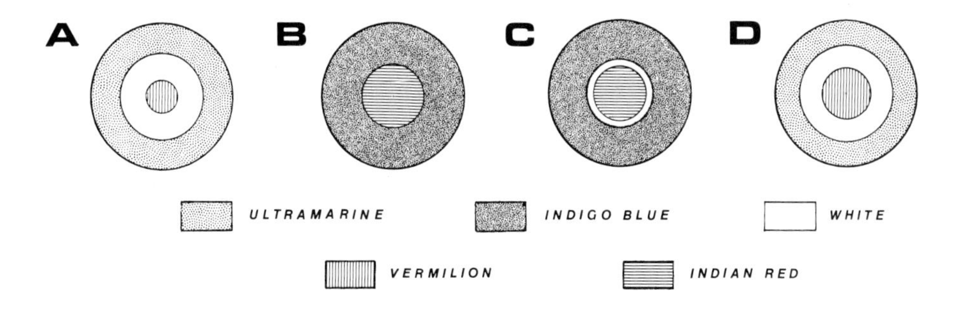

In the classification system Arabic numbers following the letter denote changes by a yellow or white surround, of varying thicknesses, to make the roundel more apparent on camouflaged surfaces—at one period early in the Second World War the British national marking was in effect a yellow ring!

Variations in presentation of the colours, bright or matt, were denoted by small Roman numerals in brackets as follows:

Code	Order of colours from the centre	Period of use
A(i)	Red, White, Blue (bright)	Late 1914-42
A(ii)	Red, White, Blue (matt)	1937-42
A1	Red, White, Blue, Yellow (matt)	1937-42
A2(i)	Red, White, Blue, Yellow (matt)	1940-2
A2(ii)	Red, White, Blue, White (bright)	1915-37
B	Red, Blue (matt)	1923-81
B1	Red, Blue, Yellow (matt)	1939 only
C	Red, White, Blue (dull)	Mid-1942-7
C1	Red, White, Blue, Yellow (dull)	Mid-1942-7
D(i)	Red, White, Blue (dull)	1947 only
D(ii)	Red, White, Blue (bright)	1947 to date

It should be appreciated that the roundel presentation on dark camouflaged upper surfaces may differ from the roundel as presented on light or 'sky' camouflaged undersurfaces, hence overlapping dates. This, the Robertson Roundel Classification System, was first introduced in 1956 to aid modellers in particular.

In the past all members of the British Empire used the RAF roundel, but later each country adopted a typical motif as the red centre to its roundel, as follows: Australia a kangaroo; Canada a maple leaf; New Zealand initially a fern leaf and later a kiwi; South Africa, changing from red to orange to represent their Dutch element, a springbok. When South Africa left the Commonwealth, the springbok was retained, but the surround became a plan of the fort of Cape Town. The Soviet satellite and communist countries in general incorporate a red star in their insignia, but exceptions are the red, white and blue segmented circle of Czechoslovakia and the red and white checks of Poland.

The United States has had more changes in the form of its national insignia than any other country. Apart from a red star used in 1916 in the Mexican campaign, a national insignia was not necessary until America joined the Allies in April 1917 and on May 19 the insignia (A) was adopted (see diagram).

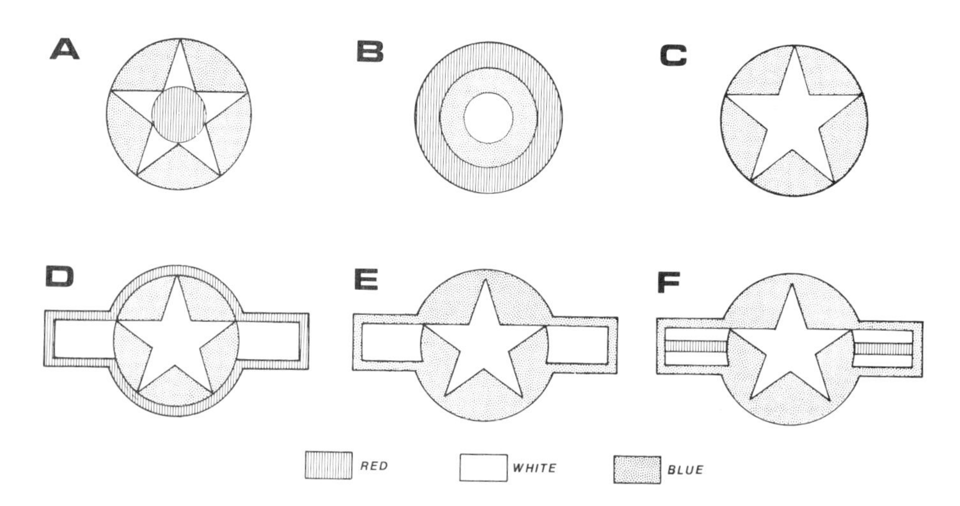

Above left *The Greek Cross* Balkankreuz, *as seen on this Messerschmitt Bf109, was used by German aircraft during 1918 and from 1935-45.*

Above right *The Iron Cross marking, first used on German aircraft from 1914 to early 1918, was re-introduced for the re-constituted Luftwaffe in 1955. It is seen on a Search and Rescue (hence the SAR) Sikorsky S-58, a type also produced in Britain as the Wessex.*

However, to conform with Allied markings on the Western Front the roundel (B) was adopted on January 11 1918. Post-war, on April 30 1919 there was a reversion to the original (A) insignia. The Japanese marking was, and still is, a red disc; to avoid confusion the red disc in the American insignia was removed mid-1942 (C). In the European theatre the fuselage roundel was given a yellow surround similar to RAF aircraft and from October the insignia was to be displayed asymmetrically on the wings—on the port upper and starboard lower only, and this regulation still appertains for all US Service aircraft. The Americans then learnt the lesson that the RFC had in 1914, that shape is more important than colour and the bands which prevail on the US aircraft insignia today were introduced in June 1943 (D). A minor change, decreed that August and taking effect from September 17, was the replacement of the red surround by a blue one (E). Finally red, to make up the colours of the national flag, was re-introduced from January 1947 in the bands (F).

Most countries of the world have national markings in the colours of their national flags or a motif indicative of their nationality. Black appears on the markings of several African countries, signifying the emergence of a black people to full sovereignty. Black also appears in practically all the insignia of Arab countries as it is the colour of the Caliphs, the defenders of the Islam faith of the past. Green was the colour symbolic of Mohammed, and this with red for fierceness in battle and white for generosity are the other favourite colours of the Arab nations. There is a need, both for the Services and the enthusiasts, for a full chart of national insignia in colour, but this is all too rarely published. There are several reasons for this, firstly a colour presentation with all the various shades correct would be expensive to produce and in black and white a colour code for such a variety of tones would be difficult to present. Then there is the timing. New nations are always emerging, other countries are changing the form of their insignia—and what of the Afghan insignia, which is always changing?

What has been achieved in this survey, is the setting out of the changes of the form of American, British and German insignia, which is an aid to correctly dating aircraft photographs and avoiding anachronisms in model marking.

Chapter 7

The identity of an aircraft

Over the years millions of aircraft have been built and over 250,000, it is estimated, are currently in flying condition. Each one has, or will have had, some form of individual identity. Normally this takes three forms.

Constructor's number—one allotted by its manufacturer at the start of the construction so that it can be traced by the makers irrespective of any other forms of identity which it may be given.

Official number—which may take the form of a military service number or civil registration letter, but in either case an identity officially allotted by the country in which it serves, and permanent for as long as the aircraft remains in the service of the country allotting the identity.

Owner's system—which varies considerably and will most likely change if ownership is transferred. This may take the form of a name or fleet number, for example, a flying school, civil or military, may use an individual letter or number on its aircraft for ease of visual identity. A service unit might use code letters, while airlines often bestow individual names.

Construction numbers

Aircraft manufacturers use their own system of numbering their products, but they are subject in Britain to official approval for governmental quality assurance. This number, also known as sequence number, appears on the airframe, stamped or affixed, to a main member of the structure—the largest bulkhead or on the mainspar centre. On American aircraft this is often affixed on the rear fuselage.

Sequence systems vary, but the most common has been that used by the Bristol Aeroplane Company in the past—a simple numbering from No 1 upwards. Fairey Aviation used such a simple numerical system but prefixed the number by F for Fairey. Unlike most other constructors, this F prefixed number was also marked on the rear of the fuselage in small characters to avoid compromising civil or Service official markings. Some other manufacturers used prefix letters which signified the manufacturer—eg, SH for Short and Harland—or a letter or number indicative of the model, eg, 18-7026 on a Piper Cub airframe implies the number of a Model 18 Super Cub. Some firms use different blocks of numbers for each different type of aircraft. Certainly there was a tendency among the more recent manufacturers to disguise their inexperience in the field by starting numbering well up the numerical scale. Even with a straightforward numerical sequence, gaps may occur owing to cancelled orders, since the numbers are usually allotted at the order stage, not on completion.

Civil registrations

The rulings, proposed in 1912 and adopted internationally by the Paris Air Convention of 1919, still apply almost in their original form: 'The nationality and registration marks borne by aircraft shall be constituted by a group of five letters; the nationality of the aircraft shall be indicated by the first letter or the first two letters of such group. A hyphen of a length equal to the width of one of the letters shall be placed immediately after the nationality letter or letters.'

It was this Paris Convention which decreed that every aircraft engaged in international navigation should have a Certificate of Airworthiness issued or rendered valid by the state whose nationality it possessed. Also that the officer in command, pilots, engineers and other members of the operating crew, should have certificates of proficiency. For these reasons states have a governmental body concerned with administrating civil aviation—or rather civil aeronautics for airships and balloons come within the aerial navigation controls. Initially in Britain, control of civil aviation was vested in the Air Ministry, and later passed to a civil ministry.

To return to the registration aspect. Provisionally, in early 1919, before the registrations had been agreed, Britain permitted the military aircraft which were being modified for civilian use to carry their military serial number marked large. The first national certificates of registration were issued in the UK in April 1919—and civil registrations were allotted in a series prefixed K, numbered from 100 upwards. Why K was chosen by Britain as a prefix was because military aircraft, then practically the only other aircraft, numbered in series prefixed by letters allotted alphabetically had only reached J. However, after the K series went to 175, the International Convention became law and Britain took up their allotted prefix G- for Great Britain.

The first British civil registrations ran from *G-EAAA* to *G-EBZZ* and then, in 1928, from *G-AAAA*. In recent years we reached *G-AZZZ* and currently allocations in sequence are lettered from *G-BAAA*. It is common now for an aircraft to be called by its last three letters and these may also be its radio call sign. However, as a recent innovation (since up to *G-ZZZZ* is available to Britain), there have been advance, out of sequence, allocations for owners who wanted something a little unusual—*G-WHIZ*!

Provisional registrations

It would be difficult for the licensing authority to approve aircraft built for experimental reasons, or a prototype, and in such cases a special permission to fly is given and temporary identity known as Class B (Provisional) Registration is used. From 1930 a letter was allotted to each firm to mark on their experimental product, with a number, from 1 upwards for each such aircraft.

During the war private venture experimental aircraft, to which neither a military serial nor civil registration was applicable, had their allotted letter prefixed to a number starting at *-0222*. Why from such an odd number? Well, it was war-time and it would give enemy intelligence something to ponder upon!

Post-war British provisional markings were changed to a number instead of a letter to denote the manufacturer, preceded by *G-* the British nationality indicating letter, so that the provisional registration could be used for ferrying

A Miles M28 is shown in September 1941.

aircraft on delivery where overflying of other nations was involved. A listing of manufacturers' letters for provisional registration appears in the appendices.

Military aircraft serials

While civil aircraft have registration letters, military aircraft are usually identified within their service by numbering systems. Britain numbers its RAF, RN and Army aircraft in a common series which started in 1911. In an initial allocation of numbers the RNAS took up *1* to *200* while the RFC followed on at *201*. The Navy soon reached *200* so that *801-1600*, later *3001-4000* and then *8001-10000*, were allotted to the RNAS while the RFC took up the intervening numbers. In early 1916, with the allocations fully subscribed, the character of serialling changed to a letter and a number. RFC aircraft were numbered from *1* to *9999* prefixed successively by letters in alphabetical sequence with certain exceptions; these being *G* reserved for captured German aircraft, and I, O, Q, U and Y which were not used to prevent confusion with numbers or other letters. The RNAS were allotted the *N* prefix out of sequence and the FAA followed this up by *S*, but by the early '30s there was only the one system for all British Service aircraft. By then all numbers started at *1000* and ended at *9999* so that, with the prefix letter, there were five characters to a serial, a ruling which still appertains. When *Z9999* was reached with the Second World War expansion programme, the five-character rule was kept by reducing numbering to *100-999* ranges with prefixes *AA, AB,* (*AC* was missed to avoid confusion with *AG*), *AD*, et seq to *AZ* and then restarting the sequence with *BA, BB,* et seq. Today allocations reach the *ZB100-999* range. What happens when *ZZ* is reached has not been revealed and probably has not yet been considered. It should be appreciated that from the *L* prefixed allocations in 1937 there were gaps in the numerical sequence for security reasons and the numbers allotted to first, second and third prototypes were spaced out. The allocations, with an explanation of the system, are contained in *British Military Aircraft Serials 1911-1979* published by Patrick Stephens Ltd.

٠ ١ ٢ ٣ ٤ ٥ ٦ ٧ ٨ ٩

0 1 2 3 4 5 6 7 8 9

Key to Arabic numbering. The photograph shows an example on a Provost trainer of the Iraqi Air Force (Percival PY4634).

The American systems have varied according to the Service, the US Navy's Bureau of Aeronautics has used simple numbering systems initially with an A prefix, stopping in the mid-'30s at *9999* when it started again at *0001* without a prefix letter. But, owing to large Second World War procurements, this series had to be abandoned early in the '40s, before *7500* was reached, to avoid duplicating the numbers of earlier aircraft still in service. A new series was started at *00001*, which continues with six digits today.

The USAAC, USAAF and USAF have used a numbering system which starts back at *1* each year and increases as the aircraft are procured throughout that fiscal year (FY). The serial number is prefixed by the last two numbers of the US FY which starts in July. Thus the first batch of aircraft ordered for the FY 1980/1 would be numbered from *81-1* onwards. While this number appears in full near the cockpit, its visible tail number uses only the last digit of the FY placed in front of the allotted number without the hyphen, thus aircraft *81-570* would have the tail number *1570*. This system was satisfactory pre-war and during the Second World War when aircraft rarely exceeded ten years in service, but now duplications are likely to occur. Another complication is that the reintroduction of camouflage has altered the way numbers have been presented.

The French Forces do not have a series of their own but use the manufacturers' construction numbers for aircraft of French origin, or previous Service identity if obtained from other countries. In general, in French numbering, prototype aircraft are numbered *01, 02, 03*, et seq, and production aircraft *1, 2, 3*, et seq, for each separate type. An exception are the Jaguars currently in service which are of two types, single-seat strike aircraft and two-seat trainers so that the numbers are prefixed accordingly *A*- and *E*-; the A for Attack version and the E for the *Ecole* (school) version.

Canada and South Africa allot numbers in batches often starting with round figures and not necessarily allocating blocks of numbers in a progressing sequence. Australian Service aircraft all bear the initial letter *A* with a number which varies according to the type of aircraft, hypenated to a series number in that type which may start at *1* or may have a higher block of numbers allotted. The RNZAF uses a numerical system prefixed *NZ*. Many of the African and Arab nations number their aircraft from *100, 200, 300*, et seq, using a different 100 range for each type of aircraft.

Individual identities

Military units and commercial organisations have their own identity systems. Shortly after the Second World War there was a move to make the individual identity of military aircraft more apparent. Apart from a general reversion in the RAF to placing serial number identity under the wings of all aircraft (removed from all but trainer and non-operational aircraft during the war), the serial numbers hitherto in 8 in digits on the rear of fuselages were marked large on the sides of aircraft in Bomber Command as individual markings. In America by the '50s USAF aircraft were sporting large 'buzz numbers'—so-called as a pilot who indulged in show-off 'buzzing' of other aircraft or locations could easily have his number recorded so that he could be traced and reprimanded. The system consisted of two letters and three numbers for each aircraft. The first letter denoted the function: *A*—Attack; *B*—Bomber; etc, as associated with type designations described elsewhere. At first *P* stood for Pursuit but, in 1948, the Americans adopted *F* for Fighter instead. The second letter was a USAF Command allocation starting at *A* for each different aircraft type, within the class. The number was the last three figures of the aircraft's serial number with the proviso that, where two aircraft had the same last three numbers, one would be suffixed *A*.

German aircraft had a *Werke Nr* (Works No) for aircraft individual identity,

An example of a buzz number is shown on a Lockheed P-80.

Forms of identity. This Fairey IIID shows its Service identity number (S1019) *and its construction number* (F756) *just discernible behind the roundel giving its national identity. Its unit identity* (40) *comes from a British Fleet marking system introduced in 1929 where Nos* 1-19 *identified fleet fighters,* 20-39 *fleet spotters,* 40-59 *reconnaissance aeroplanes,* 60-89 *torpedo bombers and* 90-98 *fleet reconnaissance aircraft* (via Glyn Owen).

but in Luftwaffe service 1933-45 there was a rigid system of fuselage code letters in which the letter after the German cross marking was the individual letter of the aircraft; often it was displayed in a different colour or was outlined. The letter following normally related to its particular *Staffel* and letters/numbers preceding the cross related to a higher formation. During the Second World War the presentation was balanced with two characters in front and two behind the cross, but for experimental aircraft there was a different coding system related to call-signs. Prior to the war, when two numbers followed the individual letter, they related to *Gruppe* and *Staffel* respectively. Various symbols on German aircraft had varying significance, but their meanings are more appropriate to a book on markings. The book *German Aircraft Markings 1939-45* by K.A. Merrick, published by Ian Allan, covers Luftwaffe markings 1933-45.

Post-war in the re-constituted Luftwaffe, code letters placed before the German cross related to a formation and a number aft of the cross to the aircraft's individual identity. However, from 1968 formation/unit code letters were replaced with a number of two digits, starting at *01*, relating to the type of aircraft, with the following numbers, aft of the cross, the aircraft's individual identity number.

There have been many different marking systems but these are more appropriate to a book on markings.

Chapter 8

Aircraft official specifications

Aircraft manufacture from conception to service is now a process taking between 10 and 20 years. This involves vast sums of money making finance, not ingenuity, the prime factor in aircraft production. For a new civil aircraft there is first the market research to detect the need. The larger the aircraft, the fewer airports of the world which it will be able to land at and the smaller the aircraft, the less its capacity. For a new light civil aircraft there is competition from the entrenched position of the American 'big three' producers in the field—Beechcraft, Cessna and Piper, each producing thousands of machines annually. A military aircraft is now rarely built as a private venture, but is produced to meet a military specification issued by a defence authority in which armament and crew to be carried and performance required will normally be stated. Firms will tender designs to meet the specification and an order for a prototype may be given to find out if in reality the type is as good as it appears on paper.

In Britain up to 1918, the Army and Navy selected from their evaluation of private designs and those of the Royal Aircraft Factory. With aircraft around £2,000 apiece, two or three prototypes were ordered. In late 1917 the Ministry of Munitions rationalised the acquisition of aircraft. HQ RFC in France and the Air Department of the Admiralty were asked to specify the types of aircraft which they required to meet their various tasks, giving performance in terms of maximum speed, ceiling and endurance, with the number of crew and armament. These specifications were circulated to industry, which was invited to submit designs. The Ministry selected the most promising and ordered prototypes.

The RFC tested their prototypes at the Aeroplane Experimental Establishment, Martlesham Heath, 1916-17, while the RNAS did so at the Isle of Grain station until the end of 1916 when landplanes were tested at Eastchurch. However, with the pending RFC/RNAS amalgamation as the RAF, it was decided late in 1917 to test all landplanes at Martlesham Heath. Post-war Martlesham Heath became the Aeroplane and Armament Experimental Establishment and later moved to Boscombe Down where it is now. The Marine Aircraft Experimental Station at the Isle of Grain, moving to Felixstowe in mid-1924 to become the Marine Aircraft Experimental Establishment, went to safer quarters in Scotland in 1940 and was closed post-war.

Returning to late 1917, according to the reports on the prototypes received from the testing stations so production orders would be placed for the successful candidates. This system remained much the same throughout the Second World

War. During 1918 up to six prototypes were ordered if tests with two different engine types were contemplated, otherwise three was normal. The financial restrictions in peace-time reduced prototypes to one or two—there was only one Spitfire and one Hurricane prototype; during the war, three prototypes became normal.

The first specifications issued in late-1917 with their contenders, given in parentheses, were:
A1a—Single-seat fighter (Armadillo, Bantam, Nighthawk, Osprey, Siskin, Snail, Snipe, Vampire, Wagtail).
A1c—Single-seat night fighter (Bobolink).
A2a—Two-seat fighter (Avro 530, Bulldog, Hippo).
A2b—Light bomber (DH10, Rhino).
A2d—Elementary training machine (Baboon).
A3b—Heavy bomber (Vimy).

Revised specifications were introduced in February 1918 as AF (Air Force) Types, later RAF Types, numbered as follows, with contending types given in parentheses:
I—Single-seat fighter, high altitude (As A1a/c except Vampire. Later Ara, Basilisk, Snark, Snapper).
Ia—Long-distance high-altitude fighter (Buzzard, Snipe Ia).
II—Single-seat fighter, ground targets (TF1 Camel, Salamander, Vampire).
IIIA—Short-distance fighter reconnaissance (Badger, Greyhound, Weasel, Whitehead project).
IIIB—Corps machine for artillery work (as IIIA).
IVA—Long-distance fighter, bombing escort changed to long-distance fighter reconnaissance (DH10, Cobham).
IVB—Long-distance photographic machine (Bourges, Cobham, Manchester, Oxford, Weasel).
V—Short-distance day bomber (DH10).
VI—Short-distance night bomber (DH10 later as IVB except Weasel).
VII—Short-distance night bomber (London, Vimy).
VIII—Long-distance night bomber (Bourges, Braemar, Cobham, DH10, Ganymede, Manchester, Oxford).
IX—Night-flying home defence machine (Vampire, later Badger, Greyhound, Weasel modified to meet spec).
X—Reserved.
XI—Long-distance gun machine (DH10, Sinaia, Martinsyde project).
XX—Ship aeroplane, single-seat fighter (Nighthawk modified as Nightjar).
XXI—Two-seat reconnaissance ship aeroplane, Type N2a (HP N2b, Panther, Sopwith B1).
XXII—Torpedo-carrying ship aeroplane, Type N1b (Blackburd, Shirl).
XXX—Large flying-boat, Type N3 (Cork, Cromarty, Valentia).
XXXI—Float seaplane for short-distance submarine patrol (Fairey IIIB, Short N2b).
XXXII—School seaplane (Sage 4c).
XXXIII—Boat seaplane, Type N4 (Atlanta).

In addition the Ministry kept a watching brief on industry for 'experimental machines possessing features likely to be of use'. A further category was 'experimental machines embodying features of general benefit to aeronautics' into which came the Bristol MR1, not suitable in itself for combat but, having an all-metal airframe, it merited attention.

In 1919 with the disbandment of the Ministry of Munitions, specifications were issued by the Director of Research (DoR) of the Air Ministry, type numbered from No 1, examples being:

DoR3—Two-seat reconnaissance (Bristol 69 project).
DoR3A—Two-seat corps reconnaissance (Bristol 69 and DH30/35 projects).
DoR4A—Long-distance bomber (Virginia).
DoR4B—Long-range bomber single-engined (Fawn).
DoR6—Single-seat shipborne fighter (Flycatcher, Plover).
DoR7A—Deck-landing reconnaissance/Spotter (replaced by Spec 3/21).
DoR8—Deck-landing torpedo bomber (replaced by Spec 3/20).
DoR9—Torpedo bomber, coastal defence (became Spec 16/22).
DoR12—Troop carrier (DH28 project).

Under Air Ministry re-organisation, specifications were issued each year from 1920 numbered by years, eg, AM Spec 37/22, meaning Air Ministry Specification No 37 of 1922. In some cases there was a role prefix letter, eg, F9/26, with significance as follows: A—Army co-operation; B—Bomber; C—transport; E—Experimental; F—Fighter; G—General purpose; GR—General Reconnaissance; M—torpedo bomber; N—Naval; O—Observation; P—bomber; PR—Photographic Reconnaissance; Q—target tug; R—Maritime Reconnaissance; S—naval tasks; T—Trainer; TX—Training Glider; X—glider.

The majority of these specifications are included in the aircraft type histories in various published works.

The specification numbering system ended in the late '40s, being replaced by Air Staff Operational or Experimental Requirements numbered with ASR, OR or ER prefixes as appropriate. The Air Ministry was absorbed into a unified Ministry of Defence controlling all three Services from 1964.

Chapter 9

The organisation of military aviation

Military aviation grew from the Army and, in our study of the subject, armies feature largely, for many military aircraft support Army field operations, directly or indirectly. So first, to give a grounding, we consider Army organisation.

Britain

The smallest unit of infantry is the platoon consisting usually of a junior officer (First or Second Lieutenant), a Sergeant, perhaps a Corporal or Lance-Corporal and up to 25 men. Four platoons make a company, traditionally of 100 men commanded by a Captain, but in practice numbers vary considerably. Several companies make up a battalion, commanded by a Major or Lieutenant-Colonel.

Up to this stage, all infantrymen are of a particular regiment. A regiment is a different kind of organisation. It is commanded by a Colonel and consists of a Depot which raises battalions on a local basis, eg, Cheshire Regiment, Royal Highland Fusiliers, Welsh Guards, etc. In times of peace a regiment may have only one active service battalion, but in time of war may raise 20. The regimental organisation is static like the garrison towns in which they are located, but the battalions they raise are available for brigading.

A brigade is composed of two or more battalions and is the smallest field formation to include Corps units such as Royal Artillery, Royal Engineers, Royal Signals, etc, to make it self-supporting and equipped for battle. Its commander is logically called a Brigadier. The title Brigadier-General used to be used and still is in the US Army and Air Force. A division is composed of two or more brigades and is commanded by a Major-General and has additional Corps units attached. In both World Wars the division was normally the largest military formation to have local association, eg, a Highland Division would be one of brigades composed of battalions from highland regiments. Brigades and regiments may be of various types, armoured, airborne, etc, and differ widely in numbers and equipment. A standard infantry division had some 18,000 men, an airborne division consisting of parachute and air landing (glider-borne) brigades, had only 12,000 men. Gliders went out in the '50s and the only post-war parachute brigade, formed with remnants of the war-time airborne divisions, the 1st and 6th, were numbered the 16th to combine the two numbers. This is but one example of the numbering of military units based on tradition, rather than sequence.

There are some collective terms commonly used which give no indication of size. A military unit can be a single company or battalion, or a battery of artillery. A soldier refers to his particular body of troops as his unit. Then there is the squad, which is a small body of troops brought together for a particular task, perhaps a fatigue party to clean up the barracks, and not necessarily of men of one particular unit. Finally there is the formation, which normally refers to the larger bodies of troops in the field such as brigades or divisions grouped under one commander. A field commander, the one in command, can be of any rank appropriate to the formation, which leads us to the higher formations.

The terms, unit, squad or formation do not appear in an Order of Battle. Currently the British Army's largest field formation is the British Corps maintained on the continent. A Corps in the past normally consisted of two or more divisions commanded by a Lieutenant-General and, unlike other units or formations, was expressed in Roman figures, eg, IV Corps. Although the British Army is the embracing name for the Army at home and overseas, in time of war there are field Armies. Such an Army consists of two or more Corps. These should not be confused with the specialist Corps, such as the Royal Army Medical Corps, or Royal Corps of Transport which are completely separate organisations for support services. Armies are known by numbers, eg, in Burma in the Second World War we had the 14th Army. In the First World War, there were five, 1st to 5th Armies on the Western Front under a Field Marshal and the whole comprised the British Expeditionary Force. In the Second World War, under General, later Field Marshal, Sir Bernard Montgomery, the British Liberation Army (BLA) went to the continent in 1944, and was known as the 21 Army Group—the number coming from the fact that it comprised of the 2nd British and 1st Canadian Armies.

To relate the Army organisation to military aviation—in Britain it was the Royal Engineers who were concerned with aeronautics in the British Army and on April 1 1911 they formed an Air Battalion with a No 1 (Airship) and No 2 (Aeroplane) Company. Such was the importance of aviation that on May 13 1912 this battalion was raised in status to a Royal Flying Corps. Since the companies had the role of scouting, until then the task of cavalry, they were made into squadrons each commanded by a Major—and so the squadron evolved as the basic flying unit. The squadron was broken down into a headquarters and three flights—A, B and C, each commanded by a Captain.

For each type of unit, then and now, an entitlement of men and equipment is laid down, called the establishment. The initial establishment of a squadron sets out the number of officers and men by ranks and trades, numbers of aircraft and MT (this stood for mechanical transport, not motor transport, for one very good reason that the RFC was allotted some Foden steam wagons) to which the unit was entitled. Its actual strength, particularly in war, varies. Squadron strengths refer to the number of men and machines actually to hand; squadron establishment refers to the men and machines it is entitled to have on hand.

From time to time establishments change. When the RFC expanded, this was achieved in two ways—increasing the number of squadrons and increasing the number of men and aircraft within each squadron (thus, 'raising the establishment', in military parlance). In 1918, squadron aircraft establishments, varying according to the role of the unit, were: fighter 25; Army co-operation 24; fighter reconnaissance, anti-submarine, torpedo-carrying, day bomber and night bomber (FE2b aircraft) 18; night bomber (Handley Page 0/400) and flying-boat 10. In the early '20s some squadrons consisted only of an HQ and a flight and were gradually brought up to the full peace-time strength of three flights each of three aircraft, with a reserve of three. Apart from their immediate establishment (IE), squadrons were authorised to hold a smaller immediate reserve (IR). A squadron's aircraft establishment would then be expressed in the form: No 99 Squadron Wellington Mk Ic 16 + 2.

Most fighter squadrons were established at 16 + 2 and they were organised on a two-flight basis of eight aircraft each. Bomber squadrons were normally 16 + 2 or 24 + 3 depending on whether accommodation allowed a two- or three-flight organisation. New bomber squadrons were often formed by taking one flight from a three-flight squadron and building it up to squadron strength.

Now a word about squadron numbering. The RFC and RNAS each allotted numbers numerically from No 1, but when the two services were amalgamated on April 1 1918, some re-numbering had to be done to prevent duplication of numbers. Thus, while former RFC squadrons retained their numbers, the RNAS squadrons had 200 added to the number, so that, for example, No 10 Squadron RNAS, known as 10 (Naval), became No 210 Squadron RAF.

This was the start of block allocations for RAF squadrons. When Special Reserve squadrons were raised in the '20s these were numbered from 500 and, when the Auxiliary Air Force was formed in 1924, squadrons were numbered from 600. Later Fleet flights were reorganised into squadrons numbered from 800 and, when a balloon barrage became necessary, balloon squadrons were numbered from 900. During the Second World War Commonwealth squadrons formed within the RAF were given numbers from 400 upwards.

On March 27 1924 the Air Ministry issued an order decreeing that, 'From the operational and administrative points of view it is considered desirable that the functional character of a unit should be embodied in its title for convenience of identification'. The order spelt out that titles of squadrons were to be in the form: No 1 (Fighter) Squadron or No 2 (Army Co-operation) Squadron, etc, in a role range which included bombing (not bomber) and communications for squadrons, and fleet fighter, fleet spotter, fleet reconnaissance, fleet torpedo, floatplane and flying-boat for the flights then in service. Later these cumbersome titles were abbreviated using the appropriate significant letters in the form No 1(F) Squadron, No 2(AC) Squadron and so on.

When war came these letters, signifying roles, were dropped for security reasons and have not been officially re-introduced. But some squadrons affect to use them now and, by being quoted in publications, it is evident that some journalists play along with them. Some, like No 70 Squadron, affect to use Roman numerals. Their authority, they claim, is the fact that it appears that way on their officially approved badge. That is so, but in the badges issued for the first 100 squadrons, a quarter have Roman numerals, and it would look silly in a listing to have a confusing mixture of Roman and Arabic. In such cases No 2 Squadron, which has the Roman II, might easily be confused with 11 (eleven). Forget the Roman—it is a fad, not official.

Aircraft markings are a vast subject on their own and require several books to cover them all. From 1916 squadrons were identified by simple devices until March 1918, when they were cancelled except for fighter squadrons. From 1924 fighter squadrons adopted colourful identity markings of checks or lines along fuselages and wings. With the introduction of camouflage immediately prior to the Second World War, code letters were allotted for squadrons and other units and remained in use until the mid-'50s. The only comprehensive listing of these

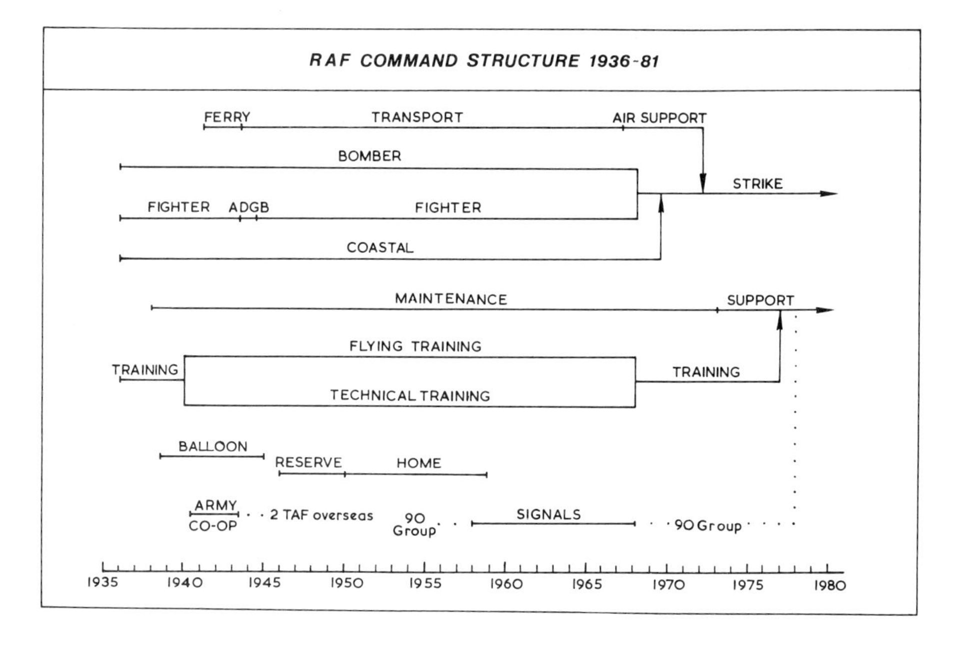

codes appeared in a series of articles in *Airfix Magazine*, published in book form in 1979 under the title *Squadron Codes 1937-56* by Michael J.F. Bowyer and John D.R. Rawlings, published by Patrick Stephens.

In the RAF a squadron's name is its number, but semi-official and nicknames have appertained. In the First World War the Nizam of Hyderabad contributed funds to pay for a squadron of DH9As and No 110 Squadron became No 110 (Hyderabad) Squadron. During the Second World War over a hundred squadrons bore titles in this manner. Then there are the completely unofficial squadron nicknames like 'The Tigers' for No 74 Squadron, which had a tiger's head squadron badge.

Now for the higher formations. A wing is a group of two or more squadrons and was first introduced in November 29 1914, commanded at Lieutenant-Colonel level—it is easy to see why, when RAF ranks were introduced, this level was made Wing Commander. Between the wars the wing organisation was not much used. At home each station, housing two or three squadrons was, in effect, a wing. The Duxford Wing in the Battle of Britain and the wings used for the following fighter sweeps over France and the Low Countries, were temporary tactical groupings.

The wings of the RFC were formed into brigades from the end of 1915 and, by 1917, there was a brigade to support each of the five British Armies on the Western Front. A brigade was logically commanded by a Brigadier, which cut out the rank of Colonel from the RFC command structure; but there were Colonels in staff positions. An RFC brigade at the Front in 1918 embraced a balloon wing and an aircraft park and transport and equipment reserves. The creation of brigades put the RFC overall command in the field up to Major-General level.

In the early RAF post-First World War re-organisation, the group replaced the brigade. With smaller numbers of men and machines it was commanded at full Colonel level, for which the RAF rank of Group Captain came into use. While the squadron has become the permanent unit of the RAF, the group has become its permanent formation. The tendency has been to upgrade the ranks for command posts and a group is now usually commanded at Air Vice-Marshal level. Groups are grouped under Commands, which have changed greatly over the years as the Command Structure Chart shows.

Germany

For an understanding of German unit organisation the reader needs to be apprised of the fact that in German the plural form is denoted by 'n' or 'en' added to the word, not by 's' as in English. Also that all German nouns start with a capital letter, not just proper nouns as in English, and that our No for number is *Nr* in German abbreviation. The basic German unit is the *Staffel* which has no direct equivalent in English for in general *Staffeln* are smaller units than squadrons.

In the First World War the first German flying unit was the *Fliegerabteilung* (*Fl Abt* for short and meaning flying section; 'en' was added in the plural form). These units for reconnaissance, photography and contact patrol were numbered from 1 upwards. *Fliegerabteilungen 'A'* (A for *Artillerie* co-operation) were numbered from 200 upwards, and units of either type on the Turkish Fronts were numbered from 300 upwards.

The German *Bomben* for bomber needs no explanation, but what we now call a fighter, the French called a chaser *(chasse)* and the Americans a pursuit, but the Germans have always called them hunters—and the German word for hunter is *Jagd.* Hence the bomber and fighter units were *Bombenstaffeln* and *Jagdstaffeln* (*Jasta* for short) respectively. *Staffeln* were grouped at times into *Geschwadern*. When in June 1917 *Jasta* 4, 6, 10 and 11 were grouped to form *Jagdgeschwader Nr 1 (JG1)*, under Manfred von Richthofen, for use along the Front as required, the word 'Circus' was soon coined by the Allies to describe this type of mobile-based fighting formation.

In the Second World War the Luftwaffe* (Air Force) was divided into *Luftflotten* (Air Fleets) of which *Luftflotten* 2, 3 and 5 opposed Fighter Command RAF in the Battle of Britain. Each *Luftflotte* consisted of one or more *Fliegerkorps* or *Fliegerdivision*, each composed of a number of *Geschwadern* which were of six main types:

Title and abbreviation	**Basic role**	**Equipment**
Kampfgeschwader (KG)	Bombing	Do17, He111, Ju88
Sturzkampfgeschwader (St KG)	Dive-bombing	Ju87 (Stukas)
Jagdgeschwader (JG)	Fighting, single-seat	Bf109, FW190
Zerstörergeschwader (ZG)	Fighting, twin-engined	Bf110
Nachtgeschwader (NJG)	Night fighting	Bf110, Ju88
Lehrgeschwader (Lebr)	Development units	Various

Each *Geschwader* was normally divided into three *Gruppen* (meaning groups) each of three *Staffeln* of nine aircraft and having an immediate reserve of another three. In addition each *Gruppe* had a *Stabs Kette* (staff section) of three operational aircraft and each *Geschwader* headquarters had a *Stabs Schwarm* (staff flight) of six operational aircraft, plus some transport and liaison aircraft, making a *Geschwader* establishment of around 100 aircraft. In each

*Though Luftwaffe is a foreign word, it is so familiar that we do not normally italicise it in English.

Geschwader the *Gruppen* and *Staffeln* were numbered consecutively; the *Gruppen* in Roman numerals I to III and the *Staffeln* in Arabic figures 1 to 9. These were used in the form II/KG40 relating to the *II Gruppe* of *Kampfgeschwader Nr 40* and 8/ZG1 would refer to the 8th *Staffel* of *Zerstörergeschwader Nr 1.*

Other German units and their abbreviations and meanings, used during the Second World War were: *Aufklärungsgruppe (Aufk/Gr)*—reconnaissance group; *Ergänzungsgruppe (Erg Gruppe)*—operational training group; *Erprobungsgruppe (Erpr Gr)*—experimental group; *Fernaufklärungsgruppe (F Gruppe)*—long-range reconnaissance group; *Heeresaufklärungsgruppe (H Gruppe)*—Army co-operation group; *Küstenfliegergruppe (KFGr)*—coastal flying group; *Nachtaufklärungsstaffel (Nasta)*—night reconnaissance unit; *Schlachtgeschwader (S* or *SG)*—ground-attack formation; *Schnellkampfgeschwader (SK* or *SKG)*—fighter-bomber formation; *Stukageschwader (StG)*—Ju87 dive-bomber formation; *Transportgeschwader (TG)*—transport formation; *Wettererkundungstaffel (Westa)*—meteorological reconnaissance unit.

Italy

The Italo-Turkish War of 1911 led to the Italians being the first to use aeroplanes in warfare. By the time Italy entered the First World War on May 24 1915, their *Corpo Aeronautico Militare* (Military Aeronautical Corps) had 89 aeroplanes in 15 *Squadriglie* (squadrons—the basic unit) which expanded to 68 by the end of the war. The Naval Air Arm, the *Aeronautica della Regia Marina*, at that time had 37 reconnaissance and nine fighter squadrons, and operated 15 airships. Post-war, on March 23 1923 the merger of the military and naval arms was approved to form the *Regia Aeronautica* (Italian Air Force) with status equal to that of the Army and Navy.

The smallest unit of the Force was the *Sezioni* (flight), but the basic unit was the *Squadriglia* (squadron) of 12IE plus 4IR aircraft in fighter squadrons and 9IE plus 3IR aircraft in bomber squadrons. Three fighter or two bomber squadrons were grouped as *Gruppi* and two or more of the latter formed *Stormi*; the singular form of these terms, it should be appreciated, is *Gruppo* and *Stormo* respectively. Formations were named according to their role—*Assaulto, Bombardmento, Osservazione* and *Transporti* need no translating, but *Caccia* for fighter is not immediately obvious until it is realised that it is similar to the French word for fighter—*Chasse*, a chaser.

The *Stormi*, commanded at full Colonel level, were allotted at home to four metropolitan air commands, 1 to 4 with headquarters at Milan, Padua, Rome and Bari respectively. A fifth command was in Tunisia, but in other areas, eg, Albania and Yugoslavia, Sardinia and Sicily, the commands had air fleet status.

After the Italian capitulation in September 1943, units of the *Regia Aeronautica* became a new Allied Air Force as the Italian Co-Belligerent Air Force. Units still under German domination were formed into the *Aviazione della RSI (Repubblica Socciale Italiana).* Post-war Italian national service aviation was re-organised as the *AMI (Aeronautica Militare Italiano)* with initially largely American, British and Italian equipment, under four regional zones. *Zone Aeree Territoriali* abbreviated to ZAT I to IV. Currently the largest tactical air formation is an air brigade *(Aerobrigata)* with assignments to the 5th Allied Tactical Air Force of NATO. The Navy and Army now have their own aviation independent of the *AMI.*

Japan

The Japanese established separate military and naval air arms before the First World War. The Army's air arm became from May 1 1925 the Military Air Corps with status equal to that of the Artillery or Engineers. It had been trained from 1919 by a French Air Mission, while the Navy's arm was trained from 1921 by a British Air Mission. The two arms differed widely in organisation and aircraft, and each had their own designating systems. Because of the difficulties of transliteration and translation the two arms were loosely referred to by the West as the JAAF (Japanese Army Air Force) and JNAF (Japanese Navy Air Force). The smallest JAAF unit was the *Shotai* corresponding to a flight with a pre-war strength of just three aircraft. The squadron equivalent was a *Chutai* of three or four *Shotai*—these were grouped into a *Sentai* (regiment) of three to four *Chutai.* A *Hikodan* (brigade) consisted of up to five *Sentai* and the next highest formation, the *Hikoshidan* (division) of two or more *Hikodan,* would have miscellaneous units, such as transport and liaison *Shotai* or *Chutai* attached. Two or more *Hikoshidan* constituted a *Kokugun* (Air Force) allotted to a theatre of operations. The JNAF had of necessity a dual system of command for fleet and land-based units.

The JAAF and JNAF were completely disbanded after 1945 and the present Japanese Air Self Defense Force (JASDF) was constituted on July 1 1954 as a part of the country's Self Defense Forces. Initial training was in the hands of the US Far East Air Force and organisation has been on American lines with the equivalent of a squadron as the basic unit.

Soviet Union

Military aeronautics in Russia dates from early in the 19th century when a balloon for observation was available to the defenders of Moscow during Napoleon's attack in 1812. Kite balloons were used at Port Arthur by Russian engineers during the Russo-Japanese War. Military aviation evolved from 1910 when a flying school was established at Gatchina near St Petersburg. An Imperial Air Service was instituted with sections subordinated to Army and Naval commands. The basic unit, an *Otryad*, consisted of six aeroplanes, later increased to ten plus two reserves. These units were grouped and allotted to Army and Naval commands. After the Revolution in 1917, aircraft used included many which had been supplied by Britain and France and Russian industry built 1,769 airframes in 1916 alone. The force was then referred to as the RKKVF *(Raboche-Krestyanskogo Krasnogo Vozclushnogo Flota)* meaning Workers' and Peasants' Red Air Fleet, and were organised into groups of *Otryad* at the revolutionary centres, which were mainly the large towns.

Under the Soviet re-organisation during the '20s Military and Naval Aviation Forces were formed with the squadron *(Eskadrilya)* the basic unit. Both forces were expanded during the Second World War when the Air Forces of the Soviet Union were divided into four main parts: Air Forces of the Red Armies, Fighter Defence Force, Long-Range Bomber Force, Air Forces of the Red Fleets.

The largest air unit in the field, was an Air Division comprised of several Air Regiments, themselves comprised of grouping squadrons. For conspicuous service by formations, their status could be raised by titling as Guards Air Regiments. Post-war, following the concept that military aviation is primarily for army support, the main body became known as FA for *Frontovaia Aviatsiya* (Frontal Aviation) organised into tactical air armies of which there are some 16 allotted to military regions in the USSR and about a quarter of these are located in Eastern Europe, based in Warsaw Pact countries and eastern parts of the Soviet Union. In all they operate some 5,000 combat aircraft and have the support of an Air Transport Force of around 1,500 fixed-wing aircraft and 3,000 helicopters, plus call on a large number of transports of Aeroflot—the state airline.

For defence of the Soviet Union there is a separate command operating fighter aircraft and missiles known as PVO for *Protivovozdushaya Oborona*, meaning protective air defence, and for strategic offence there is DA, standing for *Dal'naya Aviatsiya*, meaning long-range aviation. Naval aviation, having a divisional and regimental organisation similar to FA, has its main divisions by allocation to the five Soviet fleets.

United States

Military aviation in America was introduced by the US Army's Signal Corps which accepted its first aeroplane in 1909. This Corps raised the first Aero Squadrons, as they were called, numbered in the form 1st, 2nd, 3rd, et seq, which appertains today. After America joined the Allies in the First World War, from April 1917 control of military aviation passed to a new US Air Service and a training organisation was set up in France, Britain and Italy as well as in the USA to support an American Expeditionary Force in Europe. Post-war, in 1926, military aviation was made a Corps of the Army as the US Army Air Corps (USAAC), until June 1941 when it became US Army Air Force (USAAF), but not until June 26 1947 was the Force officially separated from the Army as an independent US Air Force (USAF).

During the Second World War the USAAF was organised into separate Air Forces for operations in the various theatres. The largest, the 8th Air Force based in Britain, was divided at peak strength into three Air Divisions. Most of the Air Forces, and the Divisions of the 8th, were broken down into two or more wings each consisting of two or more groups. This has caused some confusion because in the RAF a group is larger than a wing, but the opposite is the case in US military aviation. A group would normally be housed at one base and the average group consisted of three squadrons, the smallest USAAF/USAF operational unit. Wings and groups are designated according to their role, eg, 126th Air Refuelling Wing, 192nd Tactical Fighter Group. The USAF has a large reserve backing in its Air National Guard (ANG). The organisation of the USAF by Commands and a guide to Air Force Bases (AFB) appears each year in a special edition of the *Air Force Magazine* called the Air Force Almanac issue.

The US Navy (USN) was early in the aeronautical field dividing at first their activities into lighter- and heavier-than-air aeronautics, designated by Z and V respectively, which is reflected today in so many ways, such as US Navy squadron designations VF-1, which literally means heavier-than-air (V), fighter (F), No 1 Squadron. Similarly, on aircraft carriers the CV designations mean Carriers, Heavier-than-air aircraft. It is not only in Britain that tradition dies hard. The United States Marine Corps (USMC) also have their own aviation units and again the 'V' comes into it, eg, Squadron VMF-312 relates to heavier-than-air (V), Marine (M), Fighter (F), Squadron No 312. The form 312*th* is not used for USN/USMC squadrons, only for USAAC/USAAF/USAF numbering over the years.

Chapter 10

Services' rank structure

In any Service organisation there is a rigid chain of command in which rank structure plays an important part. The RAF first adopted its own rank names on August 4 1919, having provisionally used Army ranks from its formation on April 1 1918 when the Royal Flying Corps (RFC) and Royal Naval Air Service (RNAS) amalgamated. The RFC had used Army ranks and the RNAS their own based on Royal Navy equivalent ranks. The RAF ranks are given opposite and their other Service equivalents during the Second World War were as given below.

Royal Air Force	Royal Navy	Army
Marshal of the RAF	Admiral of the Fleet	Field Marshal
Air Chief Marshal	Admiral	General
Air Marshal	Vice-Admiral	Lieutenant-General
Air Vice-Marshal	Rear-Admiral	Major-General
Air Commodore	Commodore	Brigadier
Group Captain	Captain	Colonel
Wing Commander	Commander	Lieutenant-Colonel
Squadron Leader	Lieutenant-Commander	Major
Flight Lieutenant	Lieutenant	Captain
Flying Officer	Sub-Lieutenant	Lieutenant
Pilot Officer	Midshipman	Second Lieutenant
Warrant Officer		Warrant Officer I
	Chief Petty Officer	Warrant Officer II
Flight Sergeant		Staff Sergeant
Sergeant		Sergeant
Corporal	Petty Officer	Corporal
Leading Aircraftman	Leading Seaman	Lance Corporal
Aircraftman 1st and 2nd Class	Seaman	Private

There have been many changes in rank titles since the Second World War as the chart shows. The RAF has introduced Technician ranks which, at one period post-war, had inverted chevrons which later reverted to the form illustrated. The WAAF (Women's Auxiliary Air Force) was formed in June 1939, and became the WRAF (Women's Royal Air Force) in February 1949. A former WRAF had existed in the First World War when the Women's Royal Flying Corps became the WRAF on April 1 1918; it was disbanded precisely two years later. The senior WAAF, later WRAF, rank was Air Commandant at Air

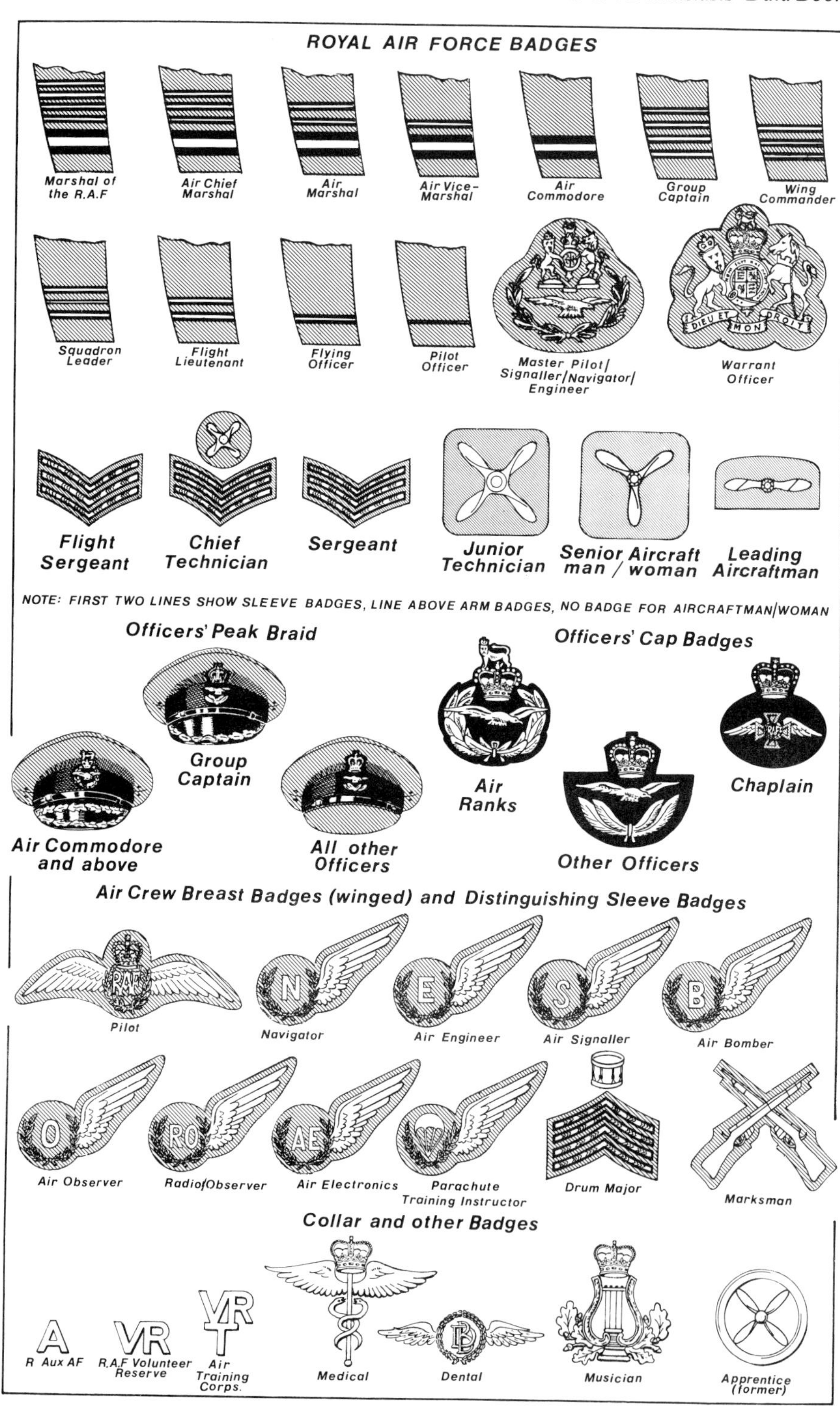
ROYAL AIR FORCE BADGES
Marshal of the R.A.F
Air Chief Marshal
Air Marshal
Air Vice-Marshal
Air Commodore
Group Captain
Wing Commander
Squadron Leader
Flight Lieutenant
Flying Officer
Pilot Officer
Master Pilot/ Signaller/Navigator/ Engineer
DIEU ET MON DROIT
Warrant Officer
Flight Sergeant
Chief Technician
Sergeant
Junior Technician
Senior Aircraft man / woman
Leading Aircraftman
NOTE: FIRST TWO LINES SHOW SLEEVE BADGES, LINE ABOVE ARM BADGES, NO BADGE FOR AIRCRAFTMAN/WOMAN
Officers' Peak Braid
Group Captain
Air Commodore and above
All other Officers
Officers' Cap Badges
Air Ranks
Other Officers
Chaplain
Air Crew Breast Badges (winged) and Distinguishing Sleeve Badges
RAF
Pilot
N
Navigator
E
Air Engineer
S
Air Signaller
B
Air Bomber
O
Air Observer
RO
Radio/Observer
AE
Air Electronics
Parachute Training Instructor
Drum Major
Marksman
Collar and other Badges
A
R Aux AF
VR
R.A.F Volunteer Reserve
VR
T
Air Training Corps.
Medical
Dental
Musician
Apprentice (former)

AMERICAN FORCES RANK BADGES / INSIGNIA

United States Air Force (Shoulder tabs unless otherwise stated)

NOTE: RANK INSIGNIA FOR U.S. ARMY AND MARINE CORPS OFFICERS SIMILAR

General of The Air Force

General

Lieutenant General

Major General

Brigadier General

Colonel

Lieutenant Colonel

Major

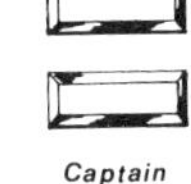
Captain

First Lieutenant

Second Lieutenant

(Blue and Silver or Gold)

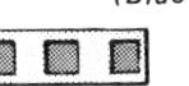

Chief Warrant Officers 2 3 4

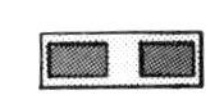
Warrant Officer 1

Sergeants (Arm only)

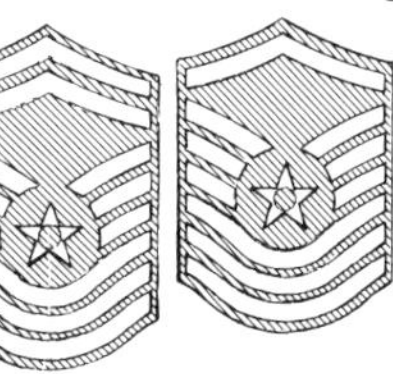
Chief Master

Senior Master

Master

Technical

Staff

Airmen (Arm only)

1st Class

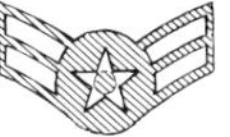
2nd Class

3rd Class

United States Navy (Sleeve badges unless othewise stated)

Fleet Admiral

Admiral

Vice Admiral

Rear Admiral

Commodore

NOTE: SHOULDER TABS BEAR ANCHOR AND STARS SIMILAR TO U.S. AIR FORCE

Captain

Commander

Lieutenant Commander

Lieutenant

Lieutenant Junior Grade

Ensign

NOTE: SHOULDER TABS BEAR SAME STAR AND STRIPES AS SLEEVE

Warrant Officers

Chief W4

Chief W3

Chief W2

W1

Chief Petty Officers (Arm only)

Master

Senior

Chief

Petty Officers (Arm only)

1st Class

2nd Class

3rd Class

Seamen (Arm only)

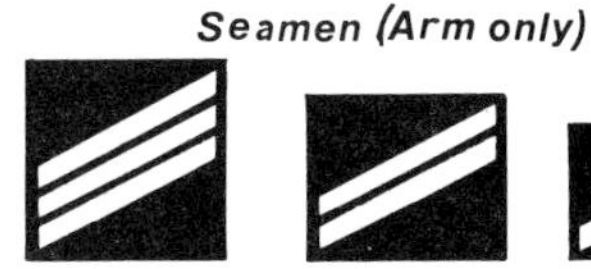
(Seaman)

Apprentice

Recruit

Commodore level and in descending order the officer rank names were: Group Officer, Wing Officer, Squadron Officer, Flight Officer, Section Officer and Assistant Section Officer. This was seen as sex discrimination, so in recent years the Force has been given the equivalent RAF rank names. The other ranks have always had RAF ranks except for Aircraftwoman instead of Aircraftman—when, one wonders, will it become Aircraftperson?

The Navy, the Senior Service because it is the oldest, would have precedence in a joint Service venture where all were of equivalent rank. It will be seen that there is considerable difference in status between a Captain and Lieutenant in the Navy and their equivalent in the Army, for this reason naval officers of these ranks are normally qualified in the form Captain RN. Unlike the other two Services the Royal Navy has a Commissioned Warrant Officer. To the layman it may come as a surprise that there is no Army rank of Sergeant Major; this is an appointment. Regimental or Company Sergeant Majors are appointed from Warrant Officer I and II ranks respectively. There are many anomalies.

The RNAS had its own officer rank names below Commodore in descending order as follows: Wing Captain, Wing Commander, Squadron Commander, Flight Lieutenant, Flight Sub-Lieutenant and Probationary Flight Sub-Lieutenant.

The American rank chart gives both Air Force and Navy as both have powerful air forces. The stars have been adopted by NATO for senior rank indications, so that a Marshal of the RAF is a five-star rank, down to Air Commodore which rates one star. It will be seen that the US Air Force ranks are Army ranks, for up to July 1947 it was an Army Air Force.

Ranks in most forces are grouped under headings. The RAF officers of Air Commodore and above are called Air Ranks, and the word Air does come into all their rank names. They, and their equivalents, are loosely called star ranks for they qualify for star plates or flags on their transport. Squadron Leader to Group Captain inclusive are classed Senior Ranks and their equivalents in the Army are known as Field Ranks. Officers below Captain in the Army are known collectively as Subalterns. Below the junior officer ranks in normal rating, but not necessarily subordinate in prestige or pay, come the Warrant Officers. While they are not commissioned, it is not correct to refer to them as NCOs (Non-Commissioned Officers) which relates to Leading Aircraftmen/women (Lance Corporal equivalent) up to Flight Sergeant (Staff or Colour Sergeant equivalent). Sergeants and above are Senior NCOs and Corporal and Leading Aircraftman (LAC)—and their equivalents—Junior NCOs. In most Service establishments Corporals and below all eat ('mess' in Service parlance) together. The so-called Sergeants Mess is officially a Warrant Officers and Sergeants Mess for Warrant Officers and Senior NCOs. Commissioned Officers of whatever rank normally have a common mess, except at some headquarters and training establishments where there are senior and junior officers messes.

Wing Commander level is now the normal rank for a Squadron Commander (which is an appointment, not a rank). In the Second World War a Wing Commander commanded bomber and multi-engined aircraft squadrons and Squadron Leaders commanded single-engined aircraft squadrons. Minimum rank for pilots has always been Sergeant, but not until 1941 was it made a minimum rank for operational aircrew. Pilots wore full wings and air gunners wore a winged round of ammunition up to 1940 when a half wing with the letters AG was introduced. Other aircrew member badges are given on the RAF badges chart, excepting the more recently introduced air quarter-masters badge with the letters QM, which both men and women NCOs may gain entitlement to wear.

Turning to other countries, many base their Air Force ranks on their Army ranks as follows:

Luftwaffe (Second World War)	**French Air Service**	**Soviet Air Force**
Reichsmarschall	—	*Marschal SU*
Generalfeldmarschall	*Marschal*	*Marschal*
General	*Générale*	*General*
Generalleutnant	*Général-Lieutenant*	*General-Leitenant*
Generalmajor	*Général-Major*	*General-Major*
Oberst	*Colonel*	*Polkownik*
Oberstleutnant	*Lieutenant-Colonel*	*Podpolkownik*
Major	*Commandant*	*Major*
Hauptmann	*Capitaine*	*Kapitan*
Oberleutnant	*Lieutenant*	*Leitenant*
Leutnant	*Sous-Lieutenant*	*Maidschy Leitenant*
Oberfeldwebel	*Adjudant-Chef*	*Starschina*
Feldwebel	*Adjudant*	*Starschij-Sergeant*
Unteroffizier	*Sergent-Chef*	*Sergeant*
Hauptgefreiter	*Sergent*	*Mladschif-Sergeant*
Obergefreiter	*Caporal*	*Jefreiter*
Gefreiter	—	—
Flieger	*Soldat*	*Ssoldat*

While the earlier table of British ranks are direct equivalents, the corresponding ranks among other countries can only be taken as approximate. For the Luftwaffe there were additional cadet *(Fähnrich)* and probationary ranks for flying personnel. An NCO pilot of the First World War was given the rank *Vizfeldwebel.*

Japanese Army Air Force	**Japanese Navy Air Force**
Ranks prefixed *'Rikugun'*	Ranks preceded *'Kaigun'*
Taisho (General)	*Taisho* (Admiral)
Chujo (Lieutenant-General)	*Chujo* (Vice-Admiral)
Shosho (Major General)	*Shosho* (Rear-Admiral)
Taisa (Colonel)	*Taisa* (Captain)
Chusa (Lieutenant-Colonel)	*Chusa* (Commander)
Shusa (Major)	*Shusa* (Lieutenant-Commander)
Taii (Captain)	*Taii* (Lieutenant)
Chui (Lieutenant)	*Chui* (Sub-Lieutenant)
Shoi (Second Lieutenant)	*Shoi* (Ensign)
Jun-1 (Warrant Officer)	*Hiko/Seibi Heisacho* (Aircrew/Ground Warrant Officer)
Socho (Sergeant Major)	*Joto Hiko/Seibi Heiso* (Superior Air/ Ground Petty Officer)
Gunso (Sergeant)	*Itto/Wito Hiko/Seibi Heiso* (1st/2nd/ Air/Ground Petty Officer)

Japanese Army Air Force
Gacho (Corporal)

Heicho (Leading Private)

Joto/Itto/Nito Hei (Superior/1st/2nd Private)

Japanese Navy Air Force
Hiko/Seibi Helcho (Leading Air/Ground Hand)
Joto Hikchei/Seibihei (Superior Air/Ground Hand)

For all technical ranks of the Japanese Navy Air Force the word *'Kogi'* followed the rank name.

Chapter 11

The language of operations

Of vital importance in any military action is meticulous planning and security. The plans may involve all three Services and detail intentions, locations, strength and timings, running at times into hundreds of sheets. For security the plans are code-named. In the Second World War the code-names of major Allied operations included: Avalanche (invading Italy at Salerno, 1943); Battleaxe (Tobruk relief, June 1941); Crossbow (measures to counter V-bombs); Diver (anti-V-1 measures); Dracula (assault on Rangoon, September 1945); Dragoon (landings in Southern France, August 1944); Dynamo (Dunkirk evacuation, May-June 1940); Fuller (*Scharnhorst* and *Gneisenau* Brest escape counter-measures); Husky (invasion of Sicily, July 1943); Market (Arnhem airborne assault, September 1944); Millennium (1,000-plus bomber raid on Cologne, May 30/31 1942); Noball (attacks on V-weapon sites 1943-44); Overlord (overall plan, invasion of Continent, June 1944); Pointblank (campaign against German fighter force and industry); Strangle (air attacks on Italian communications, March-May 1944); Torch (North African landings, November 1942); Varsity (airborne operations in support of Rhine crossing, March 1945); Zipper (projected assault on Malaya cancelled by the Japanese surrender).

The Germans similarly had code-names, examples being Barbarossa (invasion of Russia mid-1941); Hermann (attacks on Allied airfields, January 1 1945); Moonlight Sonata (Coventry bombing, November 1940); Rumpelkammer (V-1 attack on Britain); Sealion (planned invasion of Britain, 1940).

Since the date of an operation might depend on conditions of moonlight, tides, deliveries of special supplies, or when the enemy took certain measures, the precise date might not be known at the planning stage. The date was therefore referred to as D for the day—D-day. Preparatory actions, such as getting many airborne troops to the vicinity of airfields three days prior would be called D-3 (D minus 3) and supply drops required the day after landing would be D + 1 (one day after D-day). For precise timing on the day, the first assault, such as initial paratroop drops, would be H-hour, so that take-off would be around two hours previous, ie, H-2. When the date was announced to the sealed-plan holders, the units concerned would then know the date and hour to make their dispositions.

Apart from the major operation name, there were code-words for special types of operation. Bomber Command, apart from bombing operations, dropped 'Nickels' (leaflets) in 'Nickelling' (leaflet dropping) raids or went

CLOCK DIRECTIONAL REPORTING

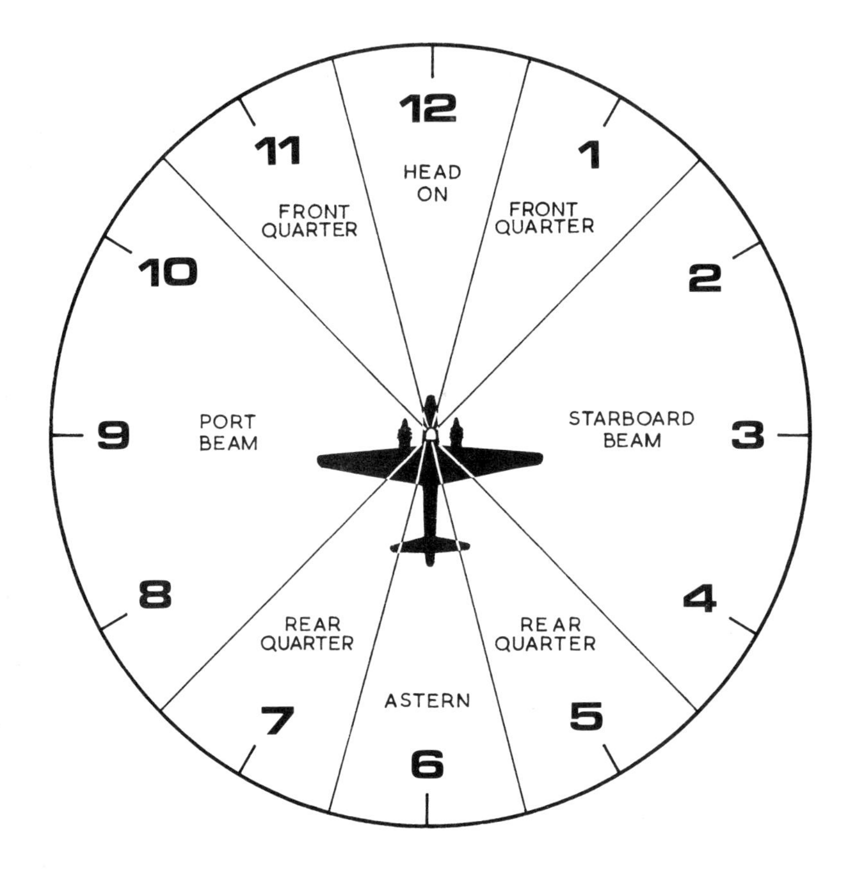

'Gardening' (mining) dropping 'Vegetables' (mines) in sea areas named regionally after flowers and vegetables. They were aided in navigation by H2S, a self-contained radar aid giving an indication of the type of territory being flown over. Other aids, depending on ground stations were 'Oboe', which gave a track signal and a time signal at bomb-dropping point for blind bombing, and 'G-H' and 'Gee' gave position fixes. For reporting and directing against attack by enemy aircraft the areas were described as shown by the chart. To confuse the enemy *Würzburg* (ground radar control system) they dropped 'Window' (metallised paper strips). For airborne and some special operations, 'Eureka' (radio beacons) were dropped on which aircraft could home. When fog affected landing, certain airfields had FIDO (Fog Investigation and Dispersal Operation) where the heat of petrol burnt in channels each side of the runway dispersed the fog as well as illuminating the runway.

A combined offensive raid by bombers and fighters from England was a 'Circus' and each one was known by a number. A purely fighter sweep was a 'Rodeo' and small-scale fighter harassing, 'Rhubarbs'. Attacks on shipping were known as 'Roadstead' operations.

Radar, initially known as RDF (Radio Direction Finding) or Radiolocation was based at AMES (Air Ministry Experimental Stations) ground stations providing CHL (Chain-High/Low) radar cover links over South-East England and was quickly extended around Britain. Night fighters were aided by acting on GCI (Ground Control Interception) radars of more limited range. These radars were adapted for bombers as 'Fishpond', to give warning of fighters.

Interceptor fighters had a language perhaps more familiar to the general public through films. They were ordered to 'scramble' (take-off) and were 'vectored' (given course to steer) at 'Angels' (height, eg, Angels 15 was 15,000 feet) to intercept 'Bandits' (enemy aircraft) and might be warned of 'Chickens' (friendly fighters) in their vicinity. They reported to control that they were 'Mattress' (below cloud), 'Popeye' (in cloud) or 'Quilt' (above cloud). Not sighting the enemy, they might be ordered to 'Orbit' (circle and search). 'Tallyho' was their report for sighting the enemy, 'Bogey' for an unidentified aircraft and 'Salvos' from a pilot implied, 'Am about to open fire, keep clear'. While these words were used over R/T, in emergencies where speech was difficult to hear, or only W/T available, Morse Code was used with significant letters for each code name, eg, PK for 'Pancake' (land, refuel and re-arm).

The word 'sortie' had a particular meaning in RAF parlance of one operation by one aircraft; so that a raid by 360 aircraft involved 360 sorties.

Training sequences have their own abbreviations and terms. In the First World War a pilot was trained first at a School of Aeronautics and first flew at one of the Reserve Squadrons (renamed Training Squadrons from May 1917) which were grouped in threes to form TDSs (Training Depot Stations) in 1918. Instructors then, and since, were trained at CFS (Central Flying School) which has maintained the standard of flying training throughout. Between the wars FTSs (Flying Training Schools) replaced TDSs. During the Second World War the sequence was ITW (Initial Training Wing) for ground instruction, EFTS (Elementary FTS), then SFTS (S for Service) in the UK or under the EATS (Empire Air Training Scheme). Owing to time spent at PDC and PRC (Personnel Despatch/Reception Centres) awaiting convoys home and receiving officer or NCO training, according to their rating, a PAFU (Pilot Advanced Flying Unit) was necessary before the pilot left flying training for an OTU (Operational Training Unit) under the operational command to which he was destined. There he was 'crewed up' with navigator, WOP/AG (wireless operator/air gunner), gunners and crew. If he was destined for multi-engined aircraft there was a further stage of HCU (Heavy Conversion Unit) where the flight engineer joined the crew. Then to operations, except for Lancaster crews which had a two-week LFS (Lancaster Finishing School) Course. From OTU crews could go on limited operations, particularly Nickelling and many OTU crews participated in Millennium.

Today SFTSs have reverted to FTSs, and OTUs have become OCUs (Operational Conversion Units).

Chapter 12

Honours and awards for air services

Apart from awards from UNO, such as for operations in Korea, or peace-keeping in Cyprus, honours and awards are made on a national basis with one great multi-national exception—the British Commonwealth. The bond of Commonwealth is vested in the Sovereign, not by governmental ties as in the days of Empire. Since honours and awards are bestowed by the Sovereign, the awards for Australian, Canadian and New Zealand forces in particular have been, and still are, basically the same as for the United Kingdom. Some other Commonwealth countries' forces have their own national systems, but are eligible for awards from the British Crown.

British Commonwealth

In Britain the rewards for public service, including those for members of the Services, are generally called honours. An official statement giving the reason for the award is called a citation and the official authority is the notification given in the *London Gazette*. In war-time notifications of awards appeared daily as deeds of gallantry were recognised, and still do at irregular intervals for acts of heroism, but there are two set periods per year for the notification of the bulk of the awards to British and Commonwealth personnel. These are the New Year's Honour List and the Queen's Birthday Honours List.

The highest awards for gallantry, taking precedence over all other honours, are the Victoria Cross (VC) and the George Cross (GC). The VC was instituted by Queen Victoria (hence its name) in 1856 during the Crimean War for acts of outstanding gallantry. The crosses, all 1,349 of them, have been hand-cast in bronze from guns captured at Sebastopol. The ribbon is plain crimson, but up to 1920 the ribbon for naval (including RNAS) personnel was blue. Recipients for deeds in the air are tabled at the end of this section. The George Cross was instituted in 1940 by King George VI to reward civilians, in particular, for similar acts of gallantry. It has been awarded to Service personnel, including women, for work in enemy territory, bomb disposal and similar deeds. This cross replaced the former Empire Gallantry Medal and some classes of the Albert Medal for life-saving which have included rescues from crashed and sinking aircraft; some former recipients of these earlier awards had them replaced by GCs but only some 200 have in all been awarded. The silver cross bears the image of St George slaying the dragon and has a plain dark blue ribbon.

Below the VC and GC come the various Orders, ranging from those of knighthood, rarely bestowed below Air Vice-Marshal and equivalent rank, down to the Order of the British Empire (OBE), not normally bestowed below Wing Commander level in the Armed Forces and Senior Executive Officer in the Civil Service. Among these is the Distinguished Service Order (DSO), established in 1886, for commissioned officers to reward meritorious action in the field in the face of the enemy. It is a reward normally bestowed for leadership with gallantry and was rarely given to junior officers whose individual acts were rewarded by decorations which come next in precedence.

Up to June 1918 officers were decorated for gallantry by two main crosses. For naval officers, including the RNAS in the First World War and RAF officers serving with the Fleet and Army officers in charge of anti-aircraft and other guns on merchant ships in the Second World War, there was the Distinguished Service Cross (DSC). For Army officers, including the RFC, there was the Military Cross (MC) which, after June 1918, was applicable only to RAF and other flying personnel for acts of gallantry on the ground. The June 1918 change came about by the introduction of the Distinguished Flying Cross (DFC) for acts of gallantry when flying on active operations against the enemy by commissioned and warrant officers and the Air Force Cross (AFC) for acts of courage and devotion in the air by officers and civilians. Included among the decorations is the Royal Red Cross for women of the nursing services. All the foregoing awards apply only to officers.

Now we come to medals. It was logical that there should have been different classes of medals for officers and men in the field for the obvious reason that the officers, as well as gallantry, had to display leadership at the same time. But this was hardly appropriate in the Second World War where the captain of an aircraft could be an officer or NCO and for the same act of gallantry and responsibility, the officer would be decorated with a cross and the NCO would be awarded a medal. When the crosses and medals for service in the air were introduced in 1918, the NCO pilot was the exception, as he is today, but in the Second World War with many NCO pilots, some with officer crew members, this disparity was rather like a class distinction. Which makes it even more remarkable that Queen Victoria should have instituted an award, the VC, superior to them all, applicable to all ranks.

For NCOs and airmen, the Distinguished Flying Medal (DFM) was instituted in June 1918 for the same circumstances as the DFC and the Air Force Medal (AFM) for conditions similar to the AFC. Up to that time non-commissioned naval personnel, including RNAS, had been eligible for the Conspicuous Gallantry Medal (CGM) introduced in 1874 and RAF personnel serving at sea remained eligible; in 1942 its scope was extended for pre-eminent bravery by Air Forces personnel flying on active operations. For bravery and resource under fire naval, including RNAS, personnel were eligible for the Distinguished Service Medal (DSM) instituted in 1914, and for which later Air Force personnel serving on board ship became eligible. RFC personnel up to mid-1918 had received the Army awards of the Distinguished Conduct Medal (DCM) introduced in 1862 and the Military Medal (MM) instituted in 1914 and which remained applicable to airmen for deeds on the ground.

Two other medals for service to the country in more general ways were, and still are, met by the award of the Medal of the British Empire (MBE) for Warrant officers and up to Squadron Leader and equivalent, and the British Empire Medal (BEM) for NCOs and seamen, privates and airmen. The ribbon is

RIBBONS OF BRITISH ORDERS, DECORATIONS & MEDALS

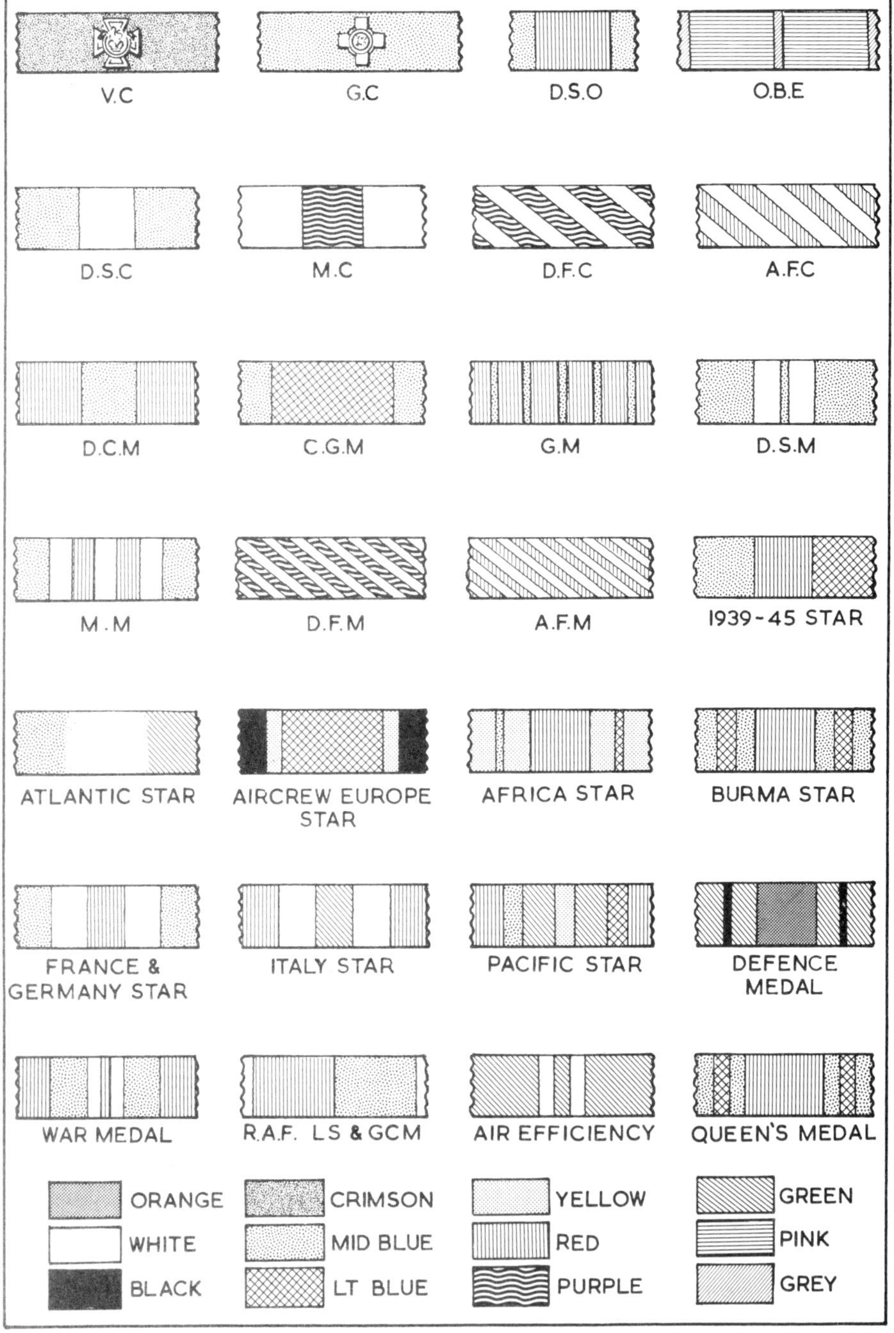

V.C
G.C
D.S.O
O.B.E

D.S.C
M.C
D.F.C
A.F.C

D.C.M
C.G.M
G.M
D.S.M

M.M
D.F.M
A.F.M
1939-45 STAR

ATLANTIC STAR
AIRCREW EUROPE STAR
AFRICA STAR
BURMA STAR

FRANCE & GERMANY STAR
ITALY STAR
PACIFIC STAR
DEFENCE MEDAL

WAR MEDAL
R.A.F. LS & GCM
AIR EFFICIENCY
QUEEN'S MEDAL

ORANGE
WHITE
BLACK
CRIMSON
MID BLUE
LT BLUE
YELLOW
RED
PURPLE
GREEN
PINK
GREY

the same colour as the OBE. It should be noted, however, that civil awards of Empire Medals do not have the central white stripe (as shown on the ribbon chart) for service recipients.

To sum up, there are only four awards exclusively for deeds of gallantry in the air, the DFC, AFC, DFM and AFM, as related. Other medals applicable to the Air Forces for deeds other than flying are the RAF Long Service and Good Conduct Medal for 18 years' exemplary service (instituted in 1919 for all ranks), the RAF Meritorious Service Award introduced in 1942 and, instituted as recently as 1953, the Queen's Medal for Champion Shots of the Air Forces. Closely associated with the air is the Royal Observer Corps Medal instituted in 1950 for officers and observers with 12 or more years 'deserving' service and the Cadet Forces Medal for officers and Warrant Officers (also instituted in that year).

Worn after the gallantry medals and rewards for particular service and before Coronation and Long Service Medals, are the war and campaign medals for service in the various theatres of war, themselves worn in order of the date of the campaign. For the First World War there were the British War Medal and Victory Medal for all who served overseas and for airmen who served operationally in the UK or were engaged on ferrying aircraft to France. Those early on the scene of operations received also the 1914-15 Star.

The campaign medals of the Second World War were also called stars and were in the shape of a six-pointed star. The 1939-45 Star was issued to personnel in the UK who completed at least 60 days with an operational unit with at least one operational sortie, or served overseas or on specified ferry or transport flights, special operations, etc. Aircrew operating over Europe were also eligible for the Air Crew Europe Star and those operating over the sea, the Atlantic Star. There were also the Africa, Burma, France and Germany, Italy and Pacific Stars issued according to the area of service. Service for 12 months in the UK, or six months if in an area under air attack, entitled Servicemen, Home Guards and Civil Defence and similar personnel to the Defence Medal and all who served for at least 28 days in the Services during the Second World War were entitled to the War Medal.

For the pre-war, inter-war and post-war years the General Service Medal, which bore a bar with the name of the area concerned, was issued for service in areas of danger such as Palestine or Malaya.

A recipient holding a distinguished service or gallantry award who later performed similar outstanding acts received a 'bar' to his decoration denoted in the form 'DSO and bar'* and the holder is sometimes referred to as a 'double DSO'. The maximum number of bars is normally three and that is extremely rarely attained. Of the highest decoration of all, the VC, only three recipients have ever been awarded a bar. Various clasps are awarded to some medals, the most well-known being an oakleaf awarded for a 'Mention in Despatches'. This clasp would be worn over the appropriate war medal.

Honours and awards are carefully regulated, not only in the rigid selection of recipients, but in the circumstances of how and when they will be worn. In general medals are worn on the left breast only on formal occasions. On evening dress at occasions where VIPs such as members of the Royal Family or even

*A bar is often indicated in printed matter such as books, newspapers, etc, by means of an asterisk—a practice which will be followed in this book.

mayors or provosts are present, miniatures (half-size representations) are worn in lieu of the originals. Each order, cross or medal, has a distinctive ribbon which is worn on everyday uniform—but not on shirts when shirt-sleeve order is the correct dress, or on overcoats. There is a strict order of precedence, and the honours related have been dealt with in that order as far as possible. This precedence conditions the order of wearing medals starting from almost the middle of the breast and moving outwards to the left. Medals with ribbons may overlap; but when ribbons alone are worn they are placed edge to edge and, being $1\frac{1}{4}$ in wide each, normally not more than five can be in a row and so, from a facing view, you read across and down in rows, in the same way that you read a book, for the order of precedence. Foreign medals, or their ribbons, are worn after those bestowed by the Crown.

No medal was issued for the Battle of Britain but a gilt rose emblem was applied to the ribbon of the 1939-45 Star of flying personnel who flew in fighter aircraft engaged in the Battle of Britain between July 10 and October 31 1940. A silver rose emblem is worn by aircrew on the appropriate campaign medal.

France

The main French awards to military personnel, including airmen, originated during the First World War. The *Croix de Guerre (CdeG)*, meaning Cross of War, was bestowed on practically all airmen who shot down an aircraft. The highest of the military decorations was the *Médaille Militaire (MM)* which was obtained by practically all French pilots achieving five or more victories. The highest honour was the combined civil/military *Légion d'Honneur (Ld'H)* which had steps from *Chevalier*, through *Officier, Commandeur, Grand Officer* to *Grand Croix*. Some airmen were promoted progressively, but even René Fonck, the top-scoring Allied ace, only reached the *Grand Officier* grade—and that post-war.

Germany

The decorations and awards to German personnel differ greatly according to the era. During the First World War there were the Imperial awards and those of the various German states. The highest decoration for bravery, the *Ordre pour le Mérite*, unlike the British VC with which it has been compared, was not awarded posthumously and was exclusively for commissioned rank. For deeds connected with the air it was awarded to the Commander and Chief of Staff of the Air Service and 59 fighter pilots, 9 observers, 5 bomber crew, 5 naval pilots including two airship commanders and a balloon observer. This order was worn at the neck from a white-edged black riband on all occasions including with field uniform. It was often referred to as the Blue Max because it was a blue cross similar to an order instituted by the Emperor Maximilian.

Another well-known German award was the Iron Cross which had several classes, the lowest of which was bestowed at unit level. Like the Blue Max this was worn with field uniform, with its position depending on its class. It was re-instituted by Hitler in September 1939 with four grades.

Under the Third Reich, a new system of awards was instituted. For outstanding deeds there was a Knights Cross *(Ritterkreuz)* of the Iron Cross worn at the neck with a riband in the Nazi colours of red, white and black. Higher grades were instituted and, although they were applicable to all Services, the first award of the Knights Cross with Oakleaves went to an airman, Werner Molders in 1940 as the first Luftwaffe pilot to gain 40 victories. During 1941 higher grades of Oakleaves with Swords and Oakleaves with Swords and Diamonds were instituted. Finally there was the *Deutsches Kreuz* (German Cross) 'in Gold' and 'in Silver' introduced in September 1941.

Russia

Under the old Imperial regime, the award for conspicuous bravery in action against the enemy was the Order of St George founded in 1769. There were four classes of the order and below them four classes of the cross and four classes of the Medal of St George. Lower classes of the cross and order were freely distributed and over two million were dispensed, so it can be assumed that every Russian pilot successful in combat received some image of St George. The civil orders of St Stanislaus and St Vladimir were also awarded for military deeds and in such cases crossed swords appeared between the limbs of the cross.

While the Soviet Union did not recognise some of the awards of Imperial Russia, the Cross of St George, as an award for bravery, could be worn with Soviet uniform. Its black and orange ribbon, signifying 'Through Darkness to Light', is used for the Soviet Decoration Order of Glory. This is one of many decorations instituted by the USSR such as the Order of Lenin and Gold Star, but the highest award is Hero of the Soviet Union. This award was instituted on April 20 1934 for the seven service and civilian pilots who rescued 111 survivors of the ice-breaker *Chelyuskin* from an ice-floe in Siberia. The Order of Lenin is automatically bestowed with the Hero award. During the Second World War some 10,000 personnel were announced as Heroes of which 2,350 were airmen, and 65 of these airmen became Twice Heroes. To Ivan Nikolia Zozhedub and Aleksandr Ivanovitch Pokryshkin, who attained 62 and 59 victories respectively, went the extreme honour of Thrice Hero of the Soviet Union.

United States

The highest American award for gallantry cited as, 'For conspicuous gallantry and intrepedity in the line of his profession at the risk of his life above and beyond the call of duty', is the Congressional Medal of Honor (MH). Ranking below the MH are the Distinguished Service Cross (DSC) and Silver Star (SS). For achievements not involving combat the Distinguished Flying Cross (DFC) and Air Medal (AM) were awarded and 41,497 and 122,745 were awarded to 8th Air Force personnel alone. For meritorious service there were issues of the Distinguished Service Medal (DSM) and Legion of Merit (LM). Where further deeds warranted recognition, Oak Leaf Clusters (OLC) were added to the medals. In addition to the awards to personnel, units were rewarded by a Distinguished Unit Citation or Meritorious Unit Plaque. American Service personnel, including attached civilians, who were wounded in action received the automatic award of the Purple Heart medal.

Victoria Cross awards for deeds in the air

'It is ordained that the Cross shall only be awarded for most conspicuous bravery, or some daring or pre-eminent act of valour or self-sacrifice of extreme devotion to duty in the presence of the enemy. It is ordained that the Cross may be awarded posthumously'—London Gazette

The following accounts are given in chronological order. It is pointed out that the dates given here are the dates of the action where appropriate, not the date of the award except in cases where a specific action is not defined. The details given are a mere brief to give the basic circumstances and in no way do justice to the full details as recorded in the citations. For full details the book *Air VCs* by Chaz Bowyer is recommended.

Lieutenant W.B. Rhodes-Moorhouse of No 2 Squadron RFC bombing in a BE2C near Courtrai on April 26 1915 was wounded by ground fire. He flew back 35 miles to base and insisted on rendering his report before going to hospital where he died next day.

Flight Sub-Lieutenant R.A.J. Warneford of No 1 Wing RNAS flying Morane Parasol *3253* on June 7 1915 attacked and destroyed the German Army Zeppelin *LZ37* over Ghent. He was killed ten days later at Buc whilst flight-testing a Henry Farman for ferrying to his unit.

Captain (later Major) L.G. Hawker of No 6 Squadron RFC flying Lewis gun armed Bristol Scout *1611*, shot down two enemy twin-seater aircraft and drove down another on July 25 1915. He was shot down in DH2 *5964* by von Richthofen, November 23 1916.

Captain J.A. Liddell MC of No 7 Squadron RFC, seriously wounded fighting a German pilot near Bruges on July 31 1915, brought his badly damaged aircraft back. He died of his wounds a month later.

Lieutenant G.S.M. Insall of No 11 Squadron RFC (with 1st Class Air Mechanic T.H. Donald) on Vickers Gunbus *5074* shot down an enemy aircraft and attacked it on the ground. His aircraft was damaged and he was forced to land just inside the French lines where repairs were made under fire and the aircraft was flown away next morning.

Squadron Commander R.B. Davies DSO of No 3 Wing RNAS landed his Nieuport Scout near Ferejik Junction to rescue a pilot, Flight Sub-Lieutenant G.F. Smylie, forced down after bombing and in imminent danger of capture by Turkish troops.

Major L.W.B. Rees commanding No 32 Squadron on July 1 1916 attacked ten German bombing aircraft in DH2 *6015* and, although wounded, continued to attack until the Germans abandoned their mission.

Captain W. Leefe Robinson of No 39 Squadron, flying BE2c *2092*, shot down the German airship *SL11* at Cuffley on the night of September 2/3 1916. He was captured when shot down in a Bristol Fighter on April 5 1917 and died in 1918 shortly after release.

Sergeant T. Mottershead DCM of No 20 Squadron RFC was flying FE2d *A39* (with Lieutenant W.E. Gower his observer) on January 7 1917, when it received a hit in the fuel tank. Although wounded and burnt, he successfully brought his aircraft down to save the observer, although he succumbed to his injuries.

Lieutenant F.H. McNamara of No 67 (Australian) Squadron RFC in Palestine on March 20 1917 landed his Martinsyde Scout, while under fire and wounded in the thigh, to rescue a BE2c pilot who had made a forced landing. After attempting to take off with the pilot the Martinsyde turned over and eventually both escaped in the BE2c.

Captain Albert Ball DSO, MC** of No 56 Squadron RFC was shot down on May 7 1917 after destroying in all 40 enemy aircraft. He was posthumously awarded the VC and *Ld'H*.

Lieutenant Colonel (later Air Marshal) W.A. Bishop DSO*, MC, DFC, *Ld'H, CdeG* of No 60 Squadron, whose favourite aircraft was Nieuport Scout

B1566, was credited with destroying 72 aircraft. He received his VC for an attack on an enemy airfield and combat in its vicinity, June 2 1917.

Major J.T.B. McCudden DSO*, MC*, MM, *CdeG* who flew in particular with No 56 Squadron RFC and was credited with 57 victories, received his VC on April 2 1918. He crashed and was killed on take-off on July 9 1918 from Auxi-le-Château in SE5A *C1126* when flying to take command of No 60 Squadron.

Second Lieutenant A.A. McLeod of No 2 Squadron RAF whilst ground strafing with guns and bombs on March 27 1918 was assailed by eight Fokker DrIs wounding him five times, yet he manoeuvred AWFK8 *B5773* to allow his observer, Lieutenant A.W. Hammond, to shoot down three. With the fuel tank set on fire, McLeod climbed on to a wing root and controlled the aircraft and dragged Hammond clear from the eventual crash-landing before collapsing himself with loss of blood.

Lieutenant A. Jerrard of No 66 Squadron RAF on the Italian Front raiding an Austrian airfield on March 30 1918 continuously attacked enemy aircraft in the air and on the ground, protecting his two comrades although wounded. He crashcd four miles from the airfield and was taken prisoner.

Major E. Mannock DSO, MC*** was posthumously awarded the VC post-war for his various actions with Nos 40, 74 and 85 Squadrons and being credited with 50 aircraft destroyed and others driven down. He was lost over the Western Front on July 26 1918.

Captain F.M.F. West MC of No 8 Squadron RAF, severely wounded on August 10 1918 when set upon by seven enemy aircraft, continued to manoeuvre his aircraft for his observer to beat off attacks. He insisted on writing his report before being taken to hospital. He retired from the RAF in 1946 as an Air Commodore with the CBE.

Captain A.W. Beauchamp-Proctor DSO, MC*, DFC, a South African of No 84 Squadron RAF flying SE5As, was awarded the VC in late December 1918 in recognition of his 54 victories and persistent ground strafing attacks.

Major W.G. Barker DSO, MC,** *Ld'H, CdeG, VM,* a Canadian, engaged in an epic air fight with German aircraft and in spite of wounds in both legs and an arm brought his Snipe *(E8102)* back for a crash landing.

Flying Officer D.E. Garland and Sergeant T. Gray, pilot and observer of a No 12 Squadron Battle, were shot down in a desperate attack to destroy bridges over the Albert Canal on May 12 1940.

Flight Lieutenant R.A.B. Learoyd pilot of Hampden *P4403* of No 49 Squadron dived through flak to attack an aqueduct on the Dortmund-Ems canal and successfully brought his badly damaged aircraft back on August 12 1940.

Flight Lieutenant J.B. Nicholson DFC flying Hurricane I *P3576* of No 249 Squadron attacked an enemy aircraft whilst his own aircraft was on fire on August 16 1940.

Sergeant J. Hannah, wireless operator/air gunner of Hampden *P1355* of No 83 Squadron, set on fire by anti-aircraft fire, September 15/16 1940, continued to fight the fire although badly burned, enabling the pilot to bring the aircraft back to base.

Flying Officer K. Campbell, piloting No 22 Squadron Beaufort *N1016* on April 6 1941, was shot down delivering a daring torpedo attack which damaged the *Gneisenau*.

Wing Commander H.I. Edwards DSO, DFC of No 105 Squadron, who had repeatedly displayed gallantry, led a force of Blenheims for a daylight attack on Bremen on July 4 1941.

Sergeant J.A. Ward of No 75 (NZ) Squadron, the second pilot of Wellington IC *L7818* on July 7 1941, climbed out on the wing in flight to extinguish a fire. He was killed in action four weeks later.

Flight Lieutenant A.S.K. Scarf making a lone raid in December 9 1941 was mortally wounded over Singora, but brought his Blenheim safely back to the Malay border.

Lieutenant Commander E. Esmonde DSO was lost February 12 1942 leading a torpedo attack by Swordfish on German capital ships passing through the Channel.

Squadron Leader J.D. Nettleton (a South African) led a daylight attack by Lancasters on the MAN diesel engine works at Augsburg, April 17 1942. He was killed in action on July 17 1943.

Flying Officer L.T. Manser, taking part in the mass raid on Cologne, on the night of May 30 1942, piloted Manchester *L7301* back to Britain in spite of its damaged condition and then sacrificed his own life in holding the aircraft steady for his crew to bale out.

Flight Sergeant R.H. Middleton, an Australian pilot serving in No 149 Squadron RAF, badly wounded and with his Stirling *(BF372)* badly damaged over Turin, on November 28 1942, flew back over the Alps to Britain with a failing aircraft which he held steady for crew members to jump, but was unable to escape himself.

Wing Commander H.G. Malcolm in *BA825* died on December 4 1942 leading a gallant attack by Blenheim Vs in the Western Desert in the face of strong enemy opposition in the air and on the ground.

Squadron Leader L.H. Trent DFC of No 487 (NZ) Squadron showed outstanding leadership and gallantry in an attack by Venturas on a power station at Amsterdam on May 3 1943 during which he was taken prisoner.

Wing Commander G.P. Gibson DSO, DFC led the famous 'dam-busting raid' the night of May 17 1943. He was killed in action the night of September 19/20 1944.

Flying Officer L.A. Trigg, a New Zealander serving in No 200 Squadron, sank a U-boat by persistent attacks in his damaged Liberator.

Flight Sergeant A.L. Aaron, although mortally wounded, brought his Stirling III *EF452* of No 218 Squadron, raiding Italy from Britain, to a safe landing in North Africa on the night of 12/13 August 1943.

Flight Lieutenant W. Reid of No 61 Squadron, piloting Lancaster III *LM360* on November 3 1943, was repeatedly wounded yet persisted in reaching his target and nursed his aircraft back to Britain.

Pilot Officer C.J. Barton of No 578 Squadron piloting Halifax III *LK797* on March 30 1944, persisted in reaching his target and was killed landing after nursing his aircraft back.

Sergeant (later Warrant Officer) N.C. Jackson of No 106 Squadron, flight engineer in Lancaster *ME669* on the night of April 26 1944, continued to deal with a fire in spite of burns and a burning parachute. He eventually baled out badly burnt and was taken prisoner.

Pilot Officer A.C. Mynarski, an air gunner on Lancaster X *KB726* of No 419 Squadron RCAF, died of injuries received the night of June 12 1944 in his gallant attempt to rescue a comrade from an aircraft being abandoned in the air.

Flight Lieutenant D.E. Hornell of No 162 Squadron RCAF on June 24 1944, engaged in a duel between his Canso and a U-boat, fighting to a finish, sinking the U-boat at the cost of his own aircraft, which had to be abandoned in the water. He died of exposure a few days later.

Flying Officer J.A. Cruikshank of No 201 Squadron continued attacking a U-boat on July 17 1944 until it was destroyed. This was in spite of his 72 wounds, loss of crew members and a badly damaged Catalina. He survived the ordeal.

Squadron Leader I.W. Bazalgette DFC, master bomber in Lancaster III *ND811* of No 635 Squadron, on August 4 1944 persisted in marking the target in spite of a burning aircraft which crashed and in which he perished.

Group Captain G.L. Cheshire DSO, DFC was awarded the VC on September 8 1944 for inter alia 'careful planning, brilliant execution and contempt for danger' after completing four operational tours.

Flight Lieutenant D.S.A. Lord of No 271 Squadron on September 19 1944 persisted in a supply drop at Arnhem with an engine on fire and under heavy fire, resulting in the crash of his Dakota in which he lost his life.

Squadron Leader R.A.M. Palmer of No 109 Squadron had completed 110 missions and was an outstanding pilot and displayed conspicuous bravery. His citation ended: 'His record of prolonged and heroic endeavour is beyond praise'. He was lost in operations December 23 1944.

Flight Sergeant G. Thompson, a wireless operator of No 9 Squadron Lancaster *PD377* damaged and on fire during a daylight attack on the Dortmund-Ems Canal, January 1 1945, gave succour to his crew members in spite of being so badly burnt that he died later of his injuries.

Captain E. Swales DFC of the SAAF, serving in No 582 Squadron on the night of February 23 1945, persisted in directing a raid, of which he was master bomber, in crippled Lancaster III *PB538*. On the return he held the aircraft steady for his crew to bale out, but it was too late to save himself.

Chapter 13

The aces

The word 'ace' has no precise meaning in aviation. Very good aerobatic pilots are loosely called aces, but the word has come in particular use to describe fighter pilots who have shot down a large number of enemy aircraft. As a standard of judging the relative achievements of pilots, victory scores have been published many times; but it should be appreciated that these are not a true criterion, for several reasons. Two of the most famous aces of the First World War were Albert Ball (British) and Max Immelmann (German), yet by victory scores they are way down the lists. They both deserve their fame for they achieved their scores before the air fighting became intense over the Western Front and long hours of patrol were flown before encountering enemy aircraft. There are also various grades of victories. Pilots were credited with 'driving down', 'driving down out of control' or 'destroying' in the official communiques. Some subsequent researchers have lumped together these reports of ascendancy over an enemy to produce the lists. Others have worked from original combat reports, without checking if the victory claimed had been officially confirmed. However, if they have been checking against enemy records not then available to the other side, their findings are likely to be more accurate than the official version.

The classic example of the official and actual versions being at variance is the claim for shooting down Manfred von Richthofen. It cannot be disputed that Captain Roy Brown was officially credited with shooting down the leading German ace on April 21 1918. In view of the overwhelming eye-witness evidence, it also cannot be disputed that his Fokker DrI triplane was seen to falter and fall from a low height while being fired upon by Australian troops, without another aircraft in the vicinity being in a position to fire upon him.

The Germans made public the victory scores of their aces as they occurred and the French gave publicity to the most outstanding of their own. It was British policy, decreed by Field Marshal Sir Douglas (later Earl) Haig, that one Corps of the Army should not be given publicity when all arms had a duty to perform to the best of their ability. And within the RFC, there was the feeling that fighter pilots should not be given publicity over those bombing or performing the prime function of the Corps in Army co-operation work.

The award of a decoration, and of the Victoria Cross in particular, did bring publicity and for this reason J.T.B. McCudden, who scored 57 victories and won the VC, was far better known than Raymond Collishaw who scored 60. The British commanders were not adverse to publicity for decorations, which applied to all arms, only to publicity for victory scores which applied to the flying services.

While there is no official meaning to the word ace, it was generally conceded by 1917, that an ace was a pilot who had downed ten or more enemy aircraft, but America, coming late into the field, used the word ace for five or more victories. As a result, to be fair, air lists when published were defined for five or more victories. For comparative purposes, the numbers of pilots obtaining five or more victories in the First World War, by nationalities were:

Nationality	No of aces	No of victories	Average score per ace
British	533	5,814	10.9
German	363	4,660	12.8
French	158	1,622	10.3
American	88	633	7.2
Italian	43	398	9.0
Austro-Hungarian	30	383	12.8
Russian	18	152	8.4
Belgian	5	71	14.2

It should be appreciated that there were many more pilots, not reaching the ace status, who shot down overall thousands of aircraft; also that Russia was out of the war from October 1917, the British score included Commonwealth pilots and the American scores include six aces who served only in the French Air Service. These figures are based on what is believed the most reliable published work on the subject to date, *Air Aces of the 1914-18 War* by Harleyford Publications, summer 1959 with a revised edition that autumn. A new in-depth aces book has been researched but has not yet been published.

It is emphasised again that the score is not the true criterion for assessing the worth of a pilot. Squadron commanders in British service, at Major level, were officially forbidden to cross the enemy lines and the patrol leaders were normally Captains or Lieutenants. Some leaders, positioning their formations for attack, would themselves go for the enemy leader, leaving the less experienced followers to those he led, rather than taking for himself an unsuspecting straggler. With two-seater aircraft, it was a pilot/gunner team; the most famous of two-seater pilots was Major A.E. McKeever DSO, MC*, a Canadian, credited with 30 victories flying a Bristol Fighter with his gunner, Sergeant (later Lieutenant) L.F. Powell, accounting for at least eight.

The leading aces of the First World War by nationality with their victory scores were: **American** 26: Captain Edward Vernon Rickenbacker MH, DSC, *Ld'H, CdeG* served with the US Air Service on the Western Front and post-war went into civil aviation. **Australian** 47: Captain R.A. Little DSO*, DSC*, *CdeG* served with the RNAS and RAF; he was killed attacking an aircraft on a night raid. The leading ace actually serving in the Australian Flying Corps was Captain A.H. Cobby, DSO, DFC**, who scored 29. **Austro-Hungarian** 40: Godwin Brumowski, a regular officer, who was killed as a passenger in an air crash near Schipol in 1937. **Belgian** 37: Lieutenant Willy Coppens de Houthulst was awarded various Belgian and French honours as well as the British DSO; when Belgium was overrun in 1940, Coppens escaped to Switzerland. **British** 73: Major Edward Mannock VC, DSO**, MC* fell on the Western Front on July 26 1918. **Canadian** 72: Lieutenant-Colonel (later Air Marshal) W.A. Bishop VC, DSO*, MC, DFC, *Ld'H, CdeG* served in the RFC and RAF. Next to him in the British listings is Lieutenant-Colonel Raymond Collishaw DSO*, DSC,

DFC, *CdeG*, the leading 'naval' ace having served in the RNAS and RAF and who became an Air Vice-Marshal in the Second World War. **French** 75: Capitaine René Paul Fonck *Ld'H, CdeG* with 28 palms, Belgian *CdeG*, British MC and MM and Russian George Cross, was the leading Allied ace of the First World War. He remained in the French Air Force and died in retirement in 1953. **German** 80: Rittmeister (Cavalry Captain) Manfred Freiherr von Richthofen, the leading ace of the First World War, was awarded the *Ordre pour le Mérite*, Order of the Red Eagle, Order of Hohenzollern with Swords, the highest gallantry medals of Bavaria, Bremen, Coburg, Oldenburg, Hamburg, Mecklenburg and Saxony, Austrian War Cross, Bulgarian Cross for bravery, and three Turkish medals. **Italian** 34: Maggiore (Major) Francesco Baracca *VM* was lost in action on June 19 1918. **Russian** 17: Staff Captain A.A. Kazakov who, apart from Russian decorations, received the British DSO, MC, DFC and the French *Ld'H*. He was killed in a take-off stall on August 2 1919.

Victory scores were for aircraft downed, so that an airman destroying a Zeppelin with the crew of some 20 officers and men would be credited with just one victory. The destruction of kite balloons was similarly counted as an aircraft score. Although they were tethered and therefore could not manoeuvre, these balloons were normally well guarded by guns and were often regarded as more difficult and more dangerous to destroy than aeroplanes. The leading German ace on the Salonikan Front, Rudolf von Eschwege who was noted for balloon attacks, was trapped and killed by Cacquot *BMS7*, an unserviceable kite balloon with a dummy observer in a basket filled with explosive which was detonated from the ground. The leading balloon ace was Michel Coffard, a Frenchman, whose 34 victories included 28 balloons. Coppens, mentioned above, shot down 26 balloons and the German Heinrich Gontermann 18.

Captain F.R. McCall, a Canadian, and Captains J.L. Trollope and H.W. Woollett, all serving in the RFC, achieved fame by shooting down six aircraft in a single day.

Between the World Wars came the Spanish Civil War of which the leading ace was Joaquin Garcia Morato, flying for the Nationalists, who had 20 Spanish pilots, and claiming five or more victories. The war was used as a testing ground for the then fascist powers and 14 German aces started counting their scores in that war; five of these were eventually credited with over 100 victories by virtue of their more rapid scoring in the Second World War. Werner Molders, the top-scoring German ace in Spain, was credited with 14 victories there. One Italian, Captain Mario Visentini, was an ace in both wars by scoring five in Spain and 12 in the Second World War. On the Government side Lieutenant Compte Rodolphe de Hemricourt de Grunne is reported as having 17 victories to which he added three in the Second World War.

The Japanese had been fighting in China since the early '30s. China's leading ace opposing them was Liu Chi Sun who scored 11 victories during 1937-41. He is also credited with sinking a Japanese warship by bombing.

How impractical a criterion of relative ability victory scores are is even more evident in the Second World War than the First. Typhoon pilots had fewer chances of combat than Spitfire pilots, yet they were doing much for the war effort in destroying locomotives, road transport, even tanks and aircraft on the ground. Then there is some disparity between shooting down aircraft on the Western Front early in the war, when most pilots were highly trained, and shooting down Russian aircraft on the Eastern Front where some Germans made claims of up to 17 victories a day. Major Heinz-Wolfgang Schnaufer credited with 126 confirmed victories, came way down on the German aces listing, but he achieved his victories at night, his victims being bombers with three to eight crew members.

Then there is the question of disparity between figures claimed at the time and those revealed by subsequent scrutiny of the records of both sides, showing claims to have been exaggerated. By nationalities the top credited scorers of the Second World War were as follows: **American** 40: Major Richard I. Bong MH, DFC, SS flying mainly P-38 Lightnings in the Pacific Theatre. He was killed on August 6 1945 whilst test-flying a P-80. **Australian** 28: Group Captain C.R. Caldwell DSO, DFC* who flew in the Western Desert. **Austrian** 258: Major Walter Nowotny, serving with the Luftwaffe, scored three on the Western and the rest on the Eastern Front. **Belgian** 11: Flight Lieutenant Vicki Ortmans DFC served with the RAF. **British** 38: Group Captain (later Air Vice-Marshal) J.E. Johnson DSO**, DFC of the RAF; the leading naval pilot was Commander C.L.G. Evans DSO, DSC of the Fleet Air Arm credited with $16\frac{1}{2}$ victories. **Bulgarian** 14: Lieutenant Stojanoff. **Canadian** 31: Squadron Leader George F. Buerling DSO, DFC, DFM* served with the RAF in Malta; he was killed when his Norseman crashed at Rome on May 20 1948. **Czech** 28: Sergeant (later Captain) Josef Frantisek received Czech, Polish and French decorations as well as a British DFM. **Danish** 9: Group Captain Kaj Birksted. **Finnish** 94: Flt Mstr E.I. Juutualainen. **French** 33: Squadron Leader Pierre Clostermann DFC* served with the RAF; the leading ace of the French Air Force 1939-40 in the Battle of France was Commandant Edmond Marin le Meslée who scored 15. **German** 352: Major Erich Hartmann, awarded the Knights Cross with Oakleaves, Swords and Diamonds, the leading ace of the Second World War, gained all but seven of his victories on the Eastern Front and was finally captured by American troops in Czechoslovakia. **Hungarian** 43: Second Lieutenant Zezsö Szentgyörgyi. **Irish** 32: Wing Commander B.E. Finucane DSO, DFC** served with the RAF and was lost on July 17 1942 trying to nurse his Spitfire back across the Channel. **Italian** 26: Maggiore Andriano Visconti, was also an ace in the Spanish Civil War. **Japanese** 103: Lieutenant Hiroyushi Nishizawa of the Navy; the top scoring Army pilot, Lieutenant-Colonel Takeo Kato, scored 58 taking in account victories in China and Mongolia. **Dutch** 5: Lieutenant-Colonel van Arkel who served with the RAF also downed 12 V-1s. **Norwegian** $16\frac{1}{2}$: Flight Lieutenant Svein Heglund served with the RAF. **Polish** 36: Jan Poniatowski served with the Royal Hellenic Air Force in Greece until shot down over Thessaly in 1941. **Romanian** 60: Captain Prince Constantine Cantacuzene. **Russian** 62: Colonel Ivan N. Kozhedub; the leading female ace was Junior Lieutenant Lydia Litvak killed in action on August 1 1943 when credited with 12 victories. **South African** 41: Squadron Leader M.T. StJ. Pattle DFC* was the RAF's leading ace by score and third in this listing was another South African, Group Captain A.G. Malan DSO*, DFC* who scored 35.

The destruction of V-1 flying-bombs was not included in ace scores. The RAF pilots destroying the greatest numbers of these pilotless aircraft were: Squadron Leader J. Berry 61, Squadron Leader van Lierde (a Belgian) 40 and Wing Commander R.P. Beamont, the famous test pilot, 32.

In the Korean War the leading ace was Captain Joseph McConnell, USAF, who flew F-86 Sabres and scored 16 victories. He was killed testing a Sabre on August 25 1954.

Chapter 14

Tactics, formations and evolutions

In all air fighting the greatest asset was height over an opponent for this gave speed in diving to the attack and allowed the choice of direction. Since it is difficult to look into the sun, except with shaded glasses impairing sight, attacks were usually made from that direction to gain the element of surprise. Once battle was joined, formations broke up into jumbled fighting to which individual tactics then applied. The various tactical manoeuvres then taken by fighter pilots and air fighting in general have rarely been explained in words or diagram. Often tactics were named after a fighter pilot who used a particular manoeuvre which hitherto had no set name, but most had slang names. The flying tactics' terms used here are based on the explanations given to Allied intelligence officers to acquaint them with air combat terms in use by both American and British pilots, which pointed out that no official adoption of the term was implied.

To fly deliberately close to and flash past another aircraft or ground location, such as a control tower, was to 'buzz' it. To 'strafe' was to attack a ground target, the term being derived from the First World War German hymn of hate *Gott strafe England* meaning 'God strike Britain*'. To 'shoot up' was usually a dive to strafe with a sharp zoom (speed) climb for recovery of altitude, also called 'dusting off' or 'flat hatting' in American slang. To push the throttle fully forward and over-ride normal maximum, was called 'full boost' or 'full bore' in British slang and to 'gun' or to 'firewall' in American.

An Immelmann turn, as illustrated, was named after Max Immelmann, one of the earliest of air aces of the First World War. An exponent of the Fokker E-series monoplanes, he scored 15 victories and was killed January 16 1916, shot down from an FE2b according to British claims or suffered a structural failure according to German sources. His manoeuvre, as subsequently assumed, achieved a complete change of direction with a gain in altitude not possible with the Fokker monoplanes that he flew.

Height is gained in a high speed climbing turn which the Americans call a 'chandelle'. Another kind of turn is known both sides of the Atlantic as a 'wing over', where the high speed climbing turn is made to a stall and the nose is allowed to fall while continuing the turn; the aircraft returns in the direction from where it came at approximately the same level. The 'split S', to use the more moderate name for this turn, is a half-roll at normal speed, followed by a pull out to normal flight in the opposite direction with some loss of height. This

*For the Germans England and Britain were synonymous.

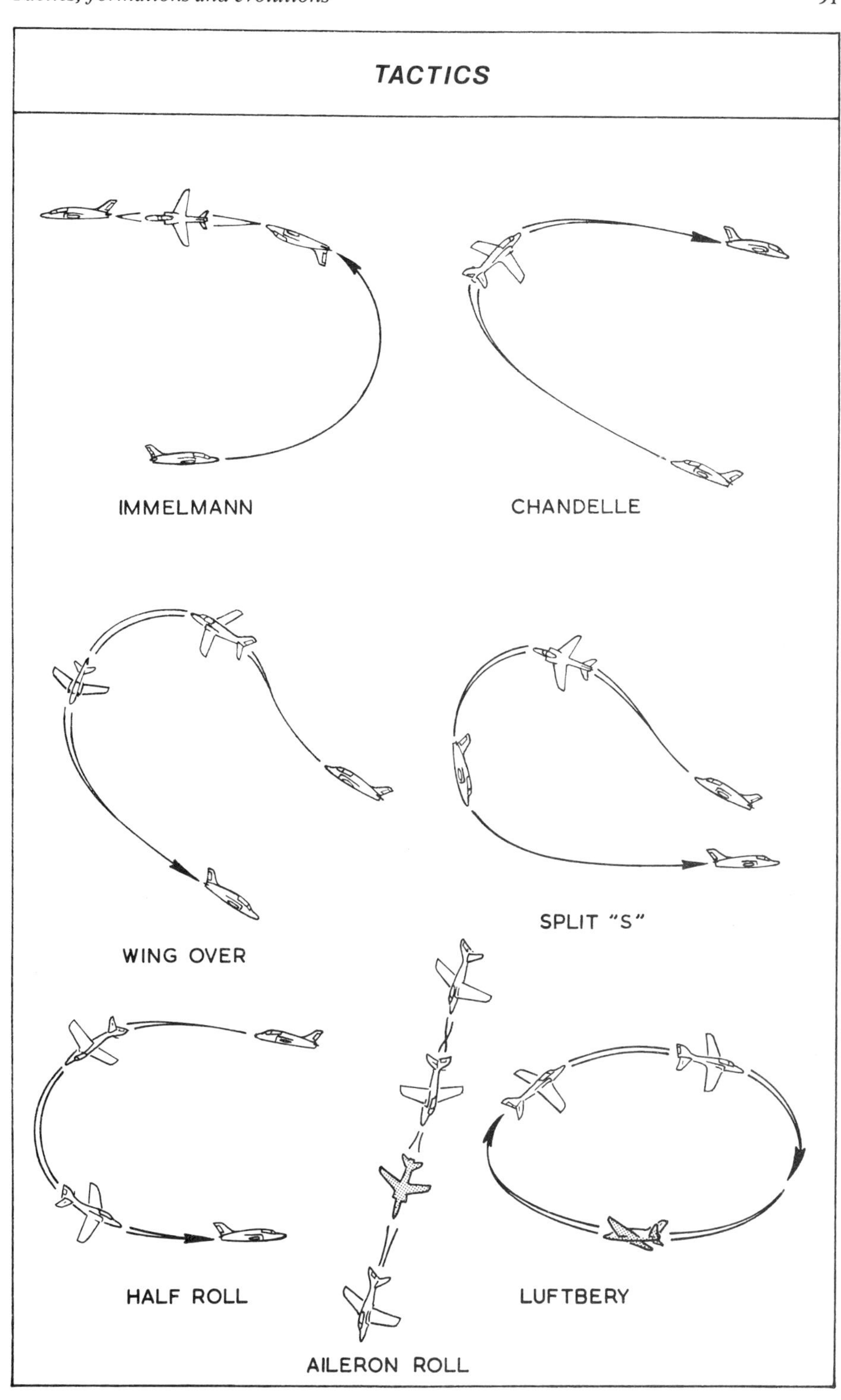
TACTICS
IMMELMANN
CHANDELLE
WING OVER
SPLIT "S"
HALF ROLL
LUFTBERY
AILERON ROLL

is similar to the so-called 'half-roll' which is the name of a turn—it should be appreciated that rolling is an aerobatic manoeuvre, not a fighting tactic. In this movement, as illustrated, the aircraft turns slowly on to its back, recovers as if from a loop, and then heads in the opposite direction to the start at a lower altitude.

The 'tail slide' shown in the evolutions is also an evasive tactic and was first called a 'whip stall' by the USAAF and is now generally known by that name. A similar tactic, with the aircraft falling off to either side was a 'stall turn' in British usage, and a 'hammer head stall' in American. To decrease speed by quick sideways movements of the tail, in fact yawing, was generally known by the Allies as 'fishtailing'. To zig-zag in direction and pitch up and down in altitude, to frustrate a fighter attack from the rear, was universally known as 'jinking', but also as 'butterflying' by the Americans.

A defensive circle of fighters to defend each others' tail was first carried out by FE2b fighter reconnaissance aircraft when heavily attacked by enemy fighters during the First World War and was colloquially known as 'ring a roses'. The Americans called this a 'Luftbery' after their First World War ace of this name. To evade an enemy fighter by diving, the aircraft was turned as if spinning down—a controlled descent known as an 'aileron roll'.

In the language of aeronautics, aircraft seen in the sky forming a regular pattern are a formation while a number of aircraft heading in the same direction without a set pattern have come to be known as a gaggle—the collective noun for geese, which is rather inappropriate since geese are one of the few species of birds capable of flying in formation. Close formations are called tight, and well-spaced formations, loose. During the Second World War the B-17 Fortresses of the US 8th Air Force flew tight formations to mass the fire of their defensive guns against enemy fighters, while the main bombers of the RAF, flying at night, took off at intervals and made their way singly to their targets to lessen the risk of collision in the dark and to avoid making concentrated targets for anti-aircraft fire. A series of formations or individual aircraft heading in the same direction is called a stream of aircraft.

The reason for including formations, evolutions and tactics here, is to help the enthusiast appreciate the terms used in books by famous airmen and to recognise some of the patterns used when watching aerobatic display teams at airshows. Apart from the basic formations, those shown in the diagram are mainly those used by that premier of aerobatic teams, The Red Arrows.

Having heard an eye-witness on TV during 1980 describe an aircraft looping, when it was rolling, led to the decision to include in this section even the basic loop. Before the Second World War, the evolutions shown were generally called stunting, but now aerobatics is the general term. A loop in a small circle is called tight, and in a large circle loose. A loop the other way, an outside loop, is called a 'bunt' and is far more difficult to perform. From time to time the RAF have issued instructions that it was a manoeuvre never to be attempted. It was first tried in 1913 by Adolphe Pegoud, a Frenchman considered the father of aerobatics and who is also credited with the first normal loop.

An aircraft cannot make a flat turn like a vehicle—but even a vehicle cannot do it at high speed, which is why race-tracks have banking. The perfect example of this are motor cycles on the Wall of Death which used to be the rage at fairgrounds. To turn, an aircraft must bank or side-slip away from the intended direction of flight. The faster and tighter the turn, the more an aircraft must

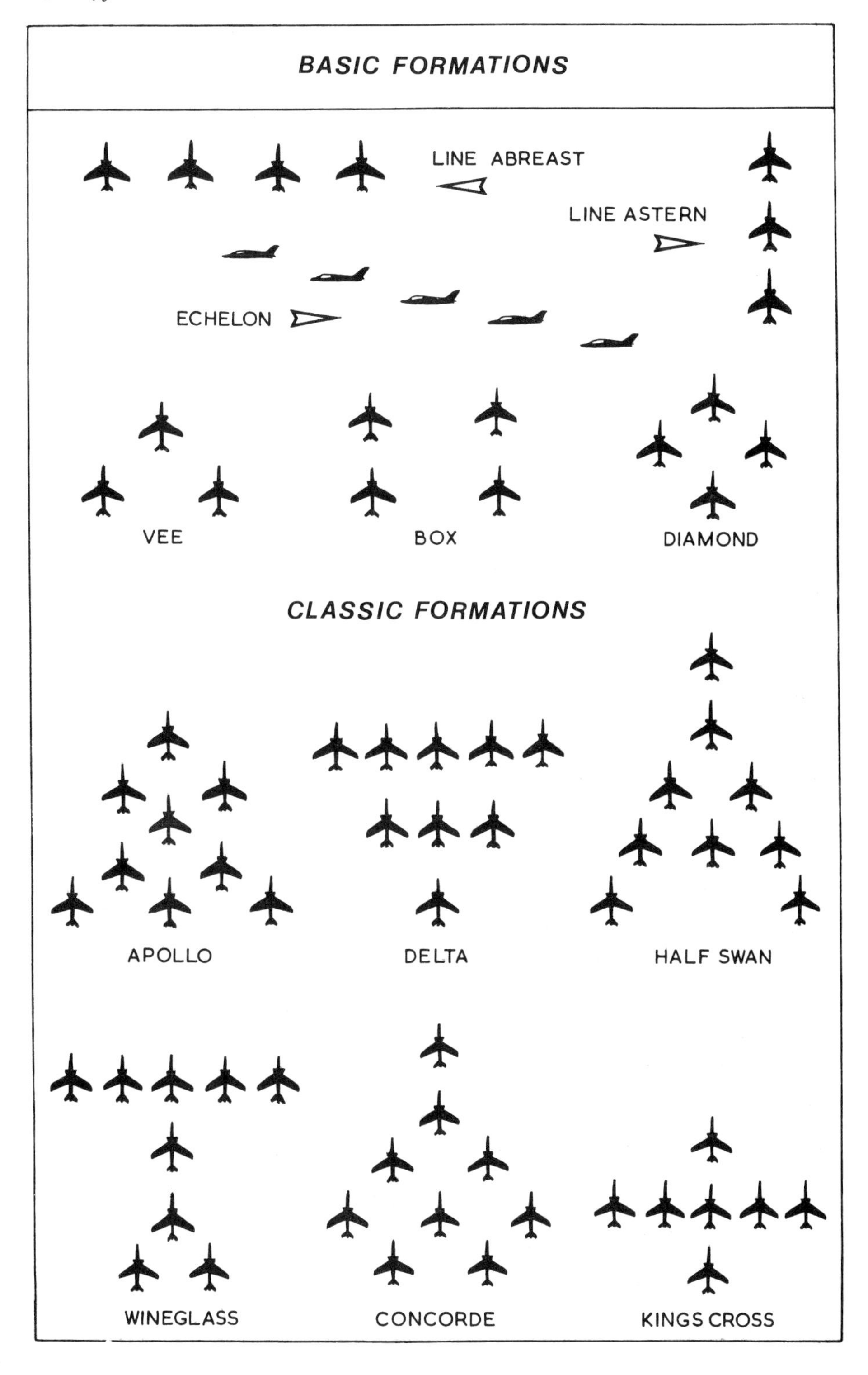
BASIC FORMATIONS
LINE ABREAST
LINE ASTERN
ECHELON
VEE
BOX
DIAMOND
CLASSIC FORMATIONS
APOLLO
DELTA
HALF SWAN
WINEGLASS
CONCORDE
KINGS CROSS

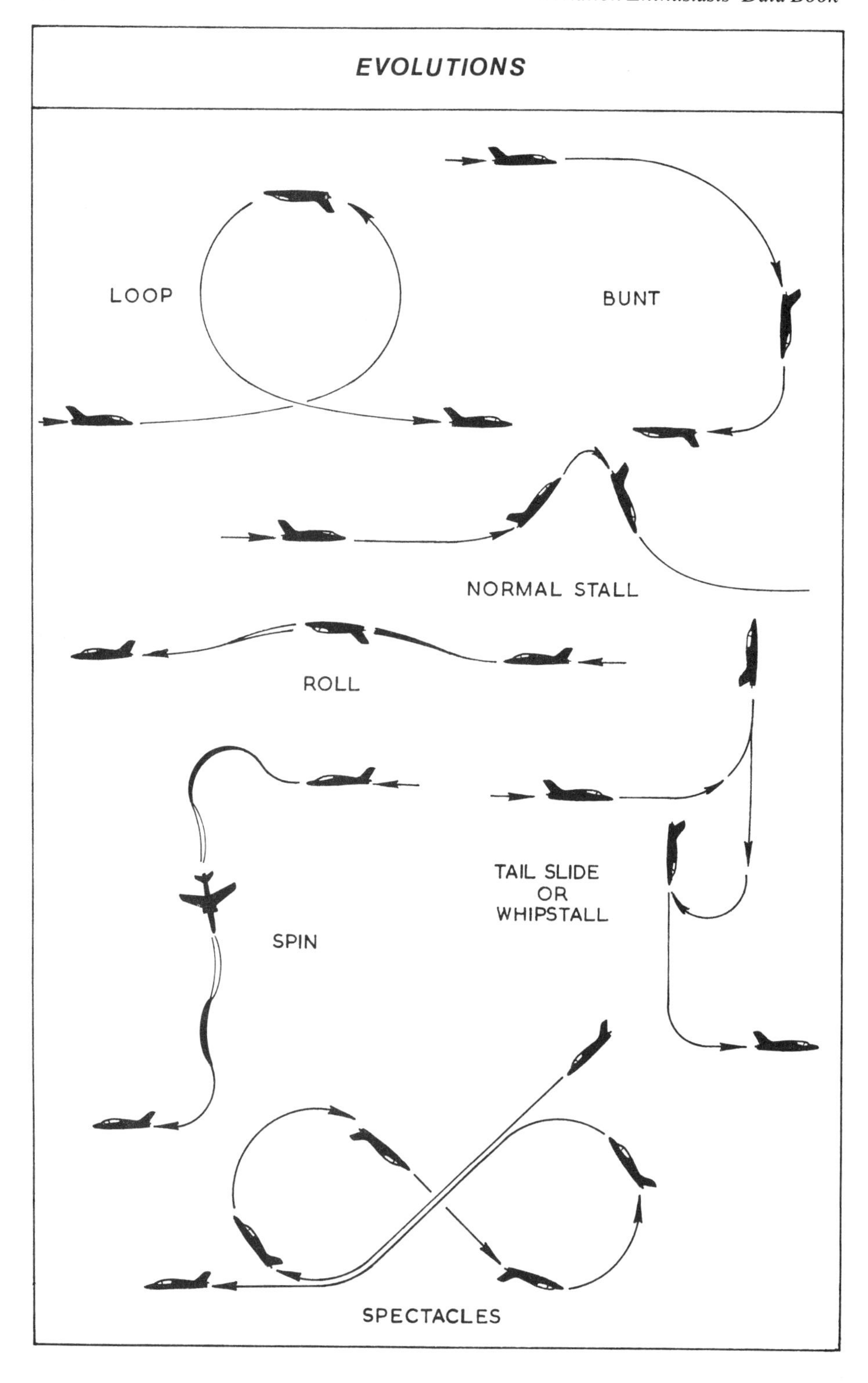
EVOLUTIONS
LOOP
BUNT
NORMAL STALL
ROLL
SPIN
TAIL SLIDE
OR
WHIPSTALL
SPECTACLES

One of the most difficult formations of all, the mirror image, demonstrated here by the 'Gemini Pair' flying Jet Provost T4s over Fylingdales radar station, Yorkshire.

bank to its maximum of 90 degrees—like the motor cycle on the Wall of Death.

In a roll the aircraft turns on its own axis or datum continuing on its same heading. A quick roll of a second or so is called a 'flick' or 'snap roll' and a slow roll, involving moments of inverted flying, a 'hesitation roll'. For a 'barrel roll' the aircraft flies a spiral. The forms that aerobatics take vary from a team presentation, when all movement must be under strict control to prevent collision, to individual demonstrations where 'crazy flying' can be indulged in allowing the aircraft to side-slip, yaw (ie, point in a slightly different position than its direction of flight—a condition that a light aircraft may experience in a high wind) or stall. An essential part of pilot training has been to induce a stall, to give the pilot sufficient confidence to recover. An aircraft stalling often goes into a spin, once (pre-1914) thought fatal, since the instinctive reaction is to pull back the control column for climb when the aircraft is falling; but first the column must go forward to stop the spin before pulling back slowly to recover. A 'flat spin' is when an aircraft is out of control turning round in swirls as it falls to earth and is not an aerobatic manoeuvre. This should not be confused with the 'falling leaf' manoeuvre of a spin with checks between the spinning in opposite ways.

Chapter 15

Air display and aerobatic teams

Flying displays in various forms have entertained the British public for 70 years. Before the First World War crowds flocked to Hendon at weekends to watch the early pioneers in aviation and after the war at the same location there were the annual displays by the RAF to thrill the public. The first was in 1920 with 'trick flying' and mock 'dog-fighting' between British and German aircraft types, but the only formation flying was a joint item by five Bristol Fighters and five Sopwith Snipes. The first five displays were called the RAF Pageant, but from 1925 this annual event became the RAF Display. Units and squadrons competed for selection; the use of smoke became a feature and, in 1930, nine Siskins performed evolutions linked together in threes by elastic cord with streamers attached. Then the Central Flying School (CFS) achieved fame for its inverted flying. The last display was the 18th in 1937. By then Empire Air Day displays at RAF aerodromes, on a Saturday in May each year 1934-9, replaced the Hendon displays. War halted further open days until September 1945 when the first of the Battle of Britain 'At Home' Day displays started with 93 stations thrown open to the public, in contrast to the two stations of some recent years.

It was 1947 before RAF display teams were again organised and, in 1950, five Vampires of No 54 Squadron were the first post-war display team to use smoke. Another two years were to elapse before team names came into vogue when the CFS formed their first jet team as 'The Meterorites'. The era of jet aircraft display teams really got under way in 1956 when the RAF fielded the 'Black Arrows', 'Black Knights' and 'Fighting Cocks' all flying front-line aircraft of the day—Hunter F4s. In the mid-'60s the RAF was forced to change from display teams by first-line units to training aircraft, but there is some consolation in the fact that this era also saw the emergence of 'The Red Arrows', the premier aerobatic display team.

Some non-Service teams have won international fame, in particular the Rothman team. A listing of some of the famous teams of the world over the years appears below in alphabetical order of team name, with brief details.

Aero Deltas, Drakens of F18 Swedish Air Force.
Air Barrels, A-4B Skyhawks of US Naval Station Glenview.
Asis of Portugal, eight Cessna T-37Cs of Portugese Air Force during the '70s.
Black Arrows, four or five Hunter F4s 1956, between seven and nine F6s 1957, nine to 16 F6s 1958-60. All of No 111 Squadron RAF, known as 'Treble-One'.
Black Diamonds, Sabre 32s of No 75 Squadron RAAF.
Black Knights, four Hunter F4s of No 54 Squadron RAF.
Blades, four Jet Provost T5s of RAF, 1970-72.

Blue Angels, four Hellcats 1946, then changed successively to Bearcats, Panthers, Cougars and Tigers with increase to six as official US Navy team. F-4J Phantoms from 1969 and changed to A-4 Skyhawks during the '70s.
Blue Bees, five Alouette IIs of Belgian Air Force, mid-'70s.
Blue Chips, two Chipmunk T10s of the RAF from the late '60s to 1974.
Blue Diamonds, 16 Hunter F6s of No 92 Squadron RAF, 1961-2, formerly **Falcons**. Also the name of team of F-86F Sabres of the Philippine Air Force.
Blue Eagles, four Sioux AH1s of the School of Army Aviation, 1968-77.
Blue Herons, five Hunter GA11s of Fleet Requirements and Development Unit (FRADU), RN, late '70s and 1980.
Blue Impulse, four F-86F Sabres of 1st Wing, Japanese Air Self Defence Force (JASDF) during '60s.
Bulldogs, two Bulldog T1s of No 3 FTS, RAF Leeming.
Carling, four Pitts Specials of Canadian civil team sponsored by Carling Breweries.
Cavillino Rampante (Prancing Horse), Vampires 1953-55, later Sabres of 4a Aerobrigata, Italian Air Force.
Diables Rouges (Red Devils), Hunter F6s replaced by six Magisters in the late '60s. Belgian Air Force part-time national team.
Diavoli Rossi (Red Devils), F-84F Thunderstreaks of 6a Aerobrigata, Italian Air Force, 1957-59.
Dolphins, three SF260s of the Belgian Air Force in the mid-'70s.
Dragonflies, four Kiowas (Bell Jet Rangers) drawn from Canadian Armed Forces training units.
Esquadrilha da Fumaca (Smoke Squadron), Harvards of the Brazilian Air Force during the '60s.
Falcons, 16 Hunter F6s renamed **Blue Diamonds** in May 1961. Also name of Jordanian team of three Pitts Specials, 1980.
Firebirds, nine Lightning F1s of No 56 Squadron RAF 1963-5.
Fighting Cocks, four Hunter F4s of No 43 Squadron in the mid-'50s.
Frecce Tricolori (three-coloured Arrows), six or seven Sabres replaced in later seasons by G91PANs as Italian Air Force national team.
Fred's Five, five Sea Vixens of 866 Squadron RN, 1962.
Gazelles, four Gazelle HT3s, of CFS, RAF, in the mid-'70s.
Gemini Pair, two Provost T4s, later T5s, of RAF Leeming.
Getti Tonanti (Thunder Jets), F-84G Thunderjets 1953-5 and F-84F Thunderstreaks 1959-60 of the Italian Air Force.
Gin also known as **Linton Gin**, Provost T4s of RAF College of Air Warfare 1968-70. Became **Blades**.
Golden Centennaires, CT-114 Tutors of the RCAF formed to commemorate 50 years of military air training in Canada.
Golden Crown, four F-84G Thunderjets 1958, four to six Sabres in the '60s of Imperial Iranian Air Force.
Golden Hawks, four Sabre 5s of RCAF 1959-64 in gold finish with red trim.
Grasshoppers, five Alouette IIIs of the Belgian Air Force, 1980.
Karo As, six Saab 105s of the Swedish Air Force, 1980.
Lanceri Neri (Black Lancers), Sabre 4s of 2a Aerobrigata, Italian Air Force, 1958-9.
Lincolnshire Poachers, Provost T4s of RAF College, Cranwell, 1969. Became **Poachers**.

Linton Blacks, four Vampire T11s of No 1 FTS, RAF Linton, 1960.
Los Jaguares, four F-86F Sabres of Venezuelan Air Force in the early '60s.
Macaws, Provost T4s of College of Air Warfare, RAF, 1968-72.
Marksmen, Sabre 31s of No 2(F) OTU, RAAF, in the mid-'60s.
Meteorites, three Meteor T7s of CFS, RAF, 1952, was also name of Meteor F8s of No 77 Squadron RAAF.
Minute Men, F-80 Shooting Stars, later F-86A Sabres, of Colorado ANG.
Patrouille d'Etampes, three later ten, Morane 230s from 1930. Changed to Morane 225s in 1934. Reformed 1947 with 12 Stampe SV4s. Team name transferred to four Vampire FB5s of 2eme Escadre de Chasse in 1950 and to ten FB5s of 4eme the following year.
Patrouille de France, four F-84 Thunderjets of 3eme Escadre, 1953. Name transferred to four Ouragans of 2eme Escadrille 1954-5 and then to other units. Changed to Mystère IVs using four in 1957, seven in 1958, nine in 1960, five in 1963 and a change to Magisters in 1964 (see below). French Air Force national team.
Patrouille de l'Ecole de l'Air, formed with Magisters in 1957 and became, on February 10 1964, ***Patrouille de France***.
Phoenix Five, five Buccaneer S2s of 809 Squadron, FAA, 1968-9.
Poachers, four Jet Provost T4s later T5As of RAF College, 1963-76.
Red Arrows, seven Gnat T1s 1965, nine Gnat T1s from 1966 and a change to Hawk T1s from 1980. RAF permanent national team.
Red Checkers, five Harvards of CFS, RNZAF.
Red Diamonds, Sabre 31s of No 76 Squadron, RAAF, in the early '60s.
Red Pelicans, four to six Jet Provost T4s of CFS from 1962, changed to T5s in 1970. (Team existed in 1960 with four T3s, but title not used).

A British Army helicopter display team of the '70s, the Blue Eagles, demonstrate a bomb burst over Corfe Castle, Dorset (Army Public Relations).

Russian roulette—a close pass by two Gnat T1s of the Red Arrows during a display in the '70s.

Redskins, two Jet Provosts of CFS, RAF, 1959.
Rothmans, Stampe SV4s 1970, changed to five Pitts S2As 1973 to maintain a team of four. Britain's only permanent civil aerobatic team.
Rough Diamonds, five Hunter GA11s of 738 Squadron, RN, 1967-8.
Sabre Knights, F-86D Sabres of 325th Fighter Interceptor Squadron, USAF.
Sharks, four to six Gazelle HT2s of No 705 Squadron, RN, 1976-81.
Silver Stars, F-86A Sabres of 335th Squadron, USAF.
Simon's Circus, six Buccaneer S2s of 892 Squadron, RN, 1968.
Skyblazers, F-86E Sabres from 1956, changing to F-100 Super Sabres in 1958. USAF team; had operated F-84G Thunderjets as team prior to adoption of name.
Skylarks, four Chipmunk T10s of CFS, RAF, 1968-70.
Snowbirds, nine CT-114 Tutors of 431 Air Demonstration Squadron, Canadian Armed Forces. Full-time team from 1974.
Sparrowhawks, Gazelle AH1s of Army Air Corps, 1979.
Swallows, three SF260s of Belgian Air Force, 1979-81.
Swords, four Jet Provost T5s of 3 FTS, RAF Leeming.
Telstars, Vampire T33s of RAAF in the mid-'60s.
Thunderbirds, four F-84G Thunderjets from 1953, F-84 Thunderstreaks from 1955, F-100C Super Sabres from 1956, then F-105B Thunderchiefs and later F-100D Super Sabres; F-4E Phantoms from 1969, but a change to T-38 Talons in the '70s. USAF national team.
Thunder Tigers, F-86F Sabres of Chinese Nationalist Air Force.
Tigers, Lightning F1s of No 74 Squadron, 1961-2.
Tigre Bianche (White Tigers), F-84G Thunderjets of Italian Air Force in the mid-'50s.
Vikings, Saab 105s of Swedish Air Force in the late-'70s.
Vintage Pair, one Meteor T7 and one Vampire T11 of CFS, RAF, 1972-81.
Vipers, Provost T4s of No 2 FTS, RAF.
Whiskey Four, four Meteors of Fighter Pilots Course at Woendrecht, whose station call-sign was Whiskey, from 1956. Lockheed T-33As from 1956. New team formed by No 314 Squadron RNedAF, 1967, with F-84F Thunderstreaks but disbanded before demonstrating.
White Swans, F-86E Sabres of Turkish Air Force in the '60s.
Yellowhammers, Vampire FB9s of No 75 Squadron, RNZAF, in the late '60s.
Yellowjacks, four Gnat T1s, forerunners of **Red Arrows**, 1964.

Among the parachute display teams are the British Army's **Red Devils** and the RAF's **Falcons**, together with the US Army **Golden Knights** and the US Navy **Chuting Stars.**

Chapter 16

Aero engines

An aeroplane is an airframe fitted with an engine, or power plant or power unit, as some choose to call this essential component. Since the earliest days of aviation the cost of the engine has roughly matched that of the airframe, and with later, larger aircraft the airframe cost is still roughly equal to that of the cost of all its engines.

Where there are more than one engine to an airframe each engine is referred to by position or number. With twin-engined aircraft the two are referred to simply as right and left engines in America and usually port (left) and starboard (right) in Britain; the left and right being that of the pilot in facing the direction of flight. With four-engined aircraft the engines are numbered from port to starboard and their respective positions in a conventional aircraft, such as a Lockheed Hercules or Boeing 707 would be port outer, port inner, starboard inner and starboard outer for engines Nos 1-4. Where engines are in tandem, as in a Cessna Skymaster, the forward engine is the No 1, and where they are superimposed as in the case of the Lightning, the lower engine is the No 1.

Steam engines proved too heavy for aeronautical use and it was the petrol-fuelled internal combustion engine development at the turn of the century which made aviation possible. In the last 40 years a new form of propulsion arrived which now divides aero engines neatly into two types—piston and jet.

Piston engines

The piston engine has been in three main forms—in-line, radial and rotary. Many readers will have at least a rudimentary knowledge of engine principles from cars. Power is supplied by explosions in chambers (cylinders) which force down pistons on to a crankshaft to convert the movement on to a revolving shaft which can turn a propeller either directly or through gearing. In some early engines of the First World War period, such as Clerget, Le Rhône and Gnôme Monosoupape, the whole radial bank of cylinders revolved and they thereby became known as rotary engines. Between the wars radial engines reached their peak; with these the cylinders remained stationary but were arranged in a radial pattern—typical examples are the famous Bristol Mercury, Pegasus, Hercules series. They went out of favour mainly because their lack of streamlining cut down speeds, but they had the great advantage of being cooled by the passage of air without the bulky and, in air warfare, vulnerable liquid cooling systems.

This leaves us with the in-line cylinder engines in use from the earliest days of aviation to today. The cylinders are in rows (or banks as they are sometimes

A typical and widely used radial engine of the Second World War, for example on Hampdens, Sunderlands and Wellingtons, the Bristol Pegasus (Bristol 5295).

called) which may be a single upright row, or more conventionally, and like many car engines, in two inclined banks, so that when you look at them from the front they form a V. Some engines have the cylinders in banks directly opposite (horizontally opposed), sometimes called flat engines.

The power of a piston engine is measured in horsepower. In terms of energy, 1 hp of work is said to be done when 33,000 lb weight is raised one foot in one minute, which can be expressed as 550 foot/lb per second. Relating this measure of power to the aero engine is not a precise matter. Power varies according to speed of revolution. A calculation for maximum output gives the engine a rating. On early aircraft engines were rated around 40-80 hp and subsequently very light aircraft have flown with motor cycle type engines of lower power. At the end of the First World War engines in use were in the 100-250 hp range, and

A typical in-line engine with two rows of six 'V'-banked cylinders represented by the Rolls-Royce Peregrine which powered the Westland Whirlwind twin-engined fighter bomber of the Second World War (Rolls-Royce 5079D).

with the Second World War a typical engine like the Rolls-Royce Merlin was developed from 1,000 to 2,000 hp. An important factor in aero engine development has been the power to weight ratio, and in the '20s designers were achieving 1 hp for every pound of weight.

Jet engines

The gas turbine engines are now almost universally referred to as jet engines and these are in four main forms—turbojet or pure jet, turbofan, turboprop and turboshaft. In a jet engine a mixture of air and gas is compressed and ignited causing rapid expansion owing to the heat generated; but instead of bearing down on a piston it is forced through a jet pipe or nozzle to provide a thrust similar to the reaction of an inflated toy balloon from which the air is suddenly released. This is the case with the so-called pure jet, which is correctly a turbojet. A form of pure jet with an extra compressor or fan which provides more air than is required for combustion and has this ducted to mix with the exhaust gases is called a turbofan. The power of both these engines is measured by their thrust in terms of pounds (lb). With a turboprop engine the energy from the exhaust stream is used to drive a turbine, and in turn a propeller, through a reduction gear. Then there is the turboshaft engine similar to the turboprop, but instead of driving a propeller it drives a shaft through reduction gearing and is used to power ships, railway engines and, in the aeronautical world, to drive helicopter rotor shafts. The power of these engines is measured in an estimated equivalent shaft horsepower (ehp).

The jet engine has a much higher power to weight ratio than the piston engine, is much smoother running with less vibration and has simplified controls. No radiators or cooling surfaces are needed and it uses less volatile fuel. However, to put it in its place, it has high fuel consumption and manufacturing costs and because of its large intakes is more susceptible to damage by ingesting objects. The piston engine will be with us for many years yet. Both engines have their place. Piston engines and turboprop engines are limited to aircraft travelling up to around 700 mph, jet engines can go supersonic up to around 2,000 mph, ramjets could go up to 3,000 mph and for rocket engines we know no limits.

Nomenclature

Britain

Early aero engines were described by their horsepower and maker, for example 100 hp Anzani, 120 hp Sunbeam, 260 hp Fiat, etc. With improvements to the basic type, mark numbers were introduced for major modifications to the basic type. Many of these changes resulted in increased horsepower, but the original figure, which in the first place may have been only approximate as far as true horsepower was concerned, remained the same for type naming purposes. One exception was the 130 hp Clerget, to which higher compression was introduced in 1917 and was first called the 130 Clerget LS (standing for the Longer Stroke of the piston rods) and then renamed the 140 hp Clerget.

In 1917 with the introduction of aeroplane nomenclature, a naming system was extended to aero engines using a different theme for each manufacturer, as follows with typical examples given in brackets: ABC, insects (Dragonfly, Wasp); Galloway, seas/oceans (Adriatic, Atlantic); Rolls-Royce, birds of prey (Hawk, Condor); Siddeley, animals (Puma, Tiger); Sunbeam, races or tribes (Arab, Maori); Wolseley-built Hispano engines, snakes (Adder, Viper). The names were applied retrospectively to earlier types in service still being developed affecting Rolls-Royce engine type names as follows:

1914-17 name	Actual hp	Post-1917 name
250 hp Mk I	225	Eagle I
250 hp Mk II	266	Eagle II
250 hp Mk III	284	Eagle IV
250 hp Mk IV	284	Eagle IV
275 hp Mk I	322	Eagle V
275 hp Mk II	322	Eagle VI
275 hp Mk III	325	Eagle VII
	350	Eagle VIII
190 hp Mk I	190	Falcon I
190 hp Mk II	250	Falcon II
	280	Falcon III

Names were not given to older types or French engines in service or to the RAF (Royal Aircraft Factory) engines which retained their Factory designations,

The Rolls-Royce Olympus turbojet, used by Concorde and Vulcan bombers, which was developed by the Bristol Aeroplane Company before the amalgamation with Rolls-Royce (Bristol 17609).

eg, 140 hp RAF4A. Similarly the Admiralty sponsored AR1 (Admiralty Rotary No 1) renamed BR1 after its designer, W.O. Bentley, and the following BR2 were not given names.

Post-war manufacturers continued using the official nomenclature both for service and commercial aircraft. Rolls-Royce continued with their birds of prey series and their Kestrel series was the prime power unit of the Hawker Hart and its variants. Various significant suffix letters were added to the mark number to denote the engine power standard: DR for derated, MS for medium supercharged and S for supercharged.

Engines were constantly being modified and, where a major modification was introduced, such as a change in propeller reduction gear, affecting performance and maintenance, a new number was issued. This was particularly likely to occur when a proven engine was adapted to a new type or new mark of aircraft. As the Roman designating figures were unwieldy for the higher numbers, it was decided in the middle of the Second World War that engine mark numbers over XX (20) would be in Arabic figures. Thus, for example, the Merlin XXIV became the Merlin 24. Merlins Mks 28, 29, 68 and 69 were American-built, but for further American manufacture of standard engines in great demand, such as the Mks 24 and 66, the figure 2 was placed in front of the number making the American-built versions the Mks 224 and 266.

Gas turbines followed the general system of manufacturer and type name with each manufacturer using a theme. In 1948 an abbreviated form was introduced for engine types using the initials of the firms and types:

Manufacturer	**Theme**	**Type name examples**	**Abbreviations**
Armstrong Siddeley	Snakes	Adder, Mamba, Viper	ASA, ASM, ASV
Bristol	Classics	Olympus, Orion, Proteus	B01, B0n, BPt
De Havilland	Spirits	Ghost, Goblin, Spectre	DGt, DGn, DSp
Napier	Species	Eland, Gazelle, Nomad	NE1, NG, NNm
Rolls-Royce	Rivers	Avon, Dart, Tyne	RA, RDa, RTy

It will be seen that a second letter of the type name had to be used in some instances. The first appearance of the engine would have the series number 1, in the manner RDa1 for the first Rolls-Royce Dart, a development became the

RDa2 and the production version, incidentally for the Viscount 700 airliner, the RDa3. Other series numbers were allotted for adaptations of the engine to other aircraft types. Sub-series numbers were also allotted. Using the Dart again as an example, the RDa7 was used for the Fokker Friendship and a similar power unit was projected for the AW650, so that the respective designations RDa7/1 and RDa7/2 applied. There were other complications in the system. Mamba, as will be seen in the table, was ASM, but the Double Mamba was expressed in the form ASMD; similarly a coupled Proteus would be given in the form RPtC.

The system had to go as the aero engine industry combined. In 1959 Bristol Aero Engines and Armstrong Siddeley Motors amalgamated to form Bristol Siddeley which later purchased the de Havilland Engine Company and Blackburn Engines. In 1966 Bristol Siddeley was acquired by Rolls-Royce and formed into their Bristol Engine Division at Filton and Small Engine Division at Leavesden, while their main Aero Engine Division remained at Derby. In addition the Rolls-Royce Motor Car Division at Crewe produced Continental light aircraft engines under licence from the USA. In 1971, Rolls-Royce (1971) Ltd was formed—with the British Government as the sole shareholder. As a result of this amalgamation British aero engines are now designated by type name and series number using suffix letters for minor modifications, for example Conway 550G, or using series and sub-series numbers in the form Spey 511-25. Additionally Rolls-Royce have used an RB-prefixed numbering system for their engines, in some cases instead of names. They are pertinent letters for a Rolls-Royce/Bristol amalgamation.

France

The French once led the world in aero engine design until the '20s when, indicative of changing times, the famous Gnôme-Rhône firm obtained a licence to produce Bristol engines. In general, from the earliest days, the French used a nominal horsepower figure and the firm's name to describe a particular type, eg, 70 hp Renault. As a model number, the number of cylinders was used with letter suffixes, eg, the first Hispano-Suiza 14-cylinder radial engine automatically became the 14A; when a smaller edition of the engine was produced it became the 14AA and the new engine model the 14AB. When new versions of the original direct-drive clockwise 14AA appeared, it was classified as the 14AA-00, and the anti-clockwise version the 14AA-01; then 14AA-03 for a geared version, and so on. Other French manufacturers used the number of cylinders and a single letter to denote changes.

Small suffix letters appeared after the number or letter in some designations, in the form Hispano-Suiza 12Ycrs, for example. These letters had the following significance:

bis—meaning encore—a further series with modification
crs—incorporating a 20 mm motor cannon firing through the boss
drs—clockwise rotation
frs—anti-clockwise rotation

During the Second World War the French aero engine industry was seriously disrupted, leading to a revival by an amalgamation of Gnôme-Rhône, Renault and others in 1945 to form SNECMA (Société Nationale d'Etude et de Construction de Moteurs d'Aviation), joined later by Hispano-Suiza. Their SNECMA Atar, first run in 1946, is still being developed. Various designations are used for engines produced in co-operation with General Electric of the USA and Rolls-Royce. The largest French aero engine producer is Turbomeca whose engines have names such as Astafan, Arbizan, Aztazou, for their gas turbines which include engines in co-operation with Rolls-Royce (the Adour) and SNECMA. Modification of type names are designated by Roman numeral mark numbers and modification of marks by suffix letters, eg Turbomeca Turbo IIIC.

Germany

Like Britain, Germany was largely dependent on French aero engines *(Flugmotoren)* before the First World War. These were classified by horsepower and manufacturing company, as in 80ps Gnome—the ps for *Pferdestärke*, the German for horsepower. During the First World War the German Army decreed a Service classification system for aero engines rather similar to that of their aeroplanes. The industry was conditioned to this system so it was all-embracing, including the German Navy. A letter or letters were allotted to each aero engine firm as follows: A—Adler Werke; As—Argus; BuS—Basse und Selve; Bz—Benz; BMW—Bayererische Motoren Werke; Goe—Goebel; Mb—Maybach; D—Mercedes (Daimler); Nag—Nationale Automobil Gesellschaft; O—Opel; Sh—Siemens-Halske and U—Oberursel.

The initial letter, indicative of the firm, was prefixed to a Roman numeral classifying its power in the range: O—up to 80 hp; I—81 hp to 100 hp; II—101 hp to 149 hp; III—150 hp to 200 hp; IV—201 hp to 300 hp; V—301 hp to 400 hp; VI—over 500 hp.

Thus BzIII would be a Benz engine in the 150-200 hp range (and not, as the popular misconception would have it, the third type of Benz engine). Its actual hp was 150, so that when a Benz engine of 185 hp appeared, still within the III class, it became the Benz IIIa and a later 195 hp model the BzIIIb. Other qualifying suffix letters were used to discriminate between a basic type and special versions as follows: m—geared; o—ungeared; ü—high compression; v—special high speed model. More than one prefix could relate to the one engine, for example, a special model of the DIIIa was built as the DIIIavü. In such cases the suffix letters would be given in alphabetical sequence.

When the German aero industry revived in the '30s a horsepower classification system was seen as being impractical, but letters indicative of the then current manufacturers were re-introduced, together with a number which would also indicate the manufacturer. This single digit number was the prefix to the engine type number allocation in a two-digit 01 to 99 series. The manufacturers' letters/figures were: BMW—BMW1 or 8; Junkers—JUM02 (the initials were derived from Junkers Flugzeug und Motoren Werke); Bramo—Sh3 initially (the letters originating from Siemens-Halske which Bramo took over) but changed to BMW3 when, in turn, BMW took over Bramo in 1939; Argus—As4; Hirth-Motoren—HM5; Daimler-Benz—DB6; Klockner-Humboldt-Deutz—Dz7; Bucker—M7 (this was a later allocation for the manufacturer of engines for the company's own training aircraft). The figure 9 was reserved. Thus, under the system, the first (01) Daimler-Benz (DB6) engine would be expressed as the DB601 and subsequent models were numbered DB602, DB603, et seq. Modifications to the basic types were indicated by suffix letters allotted alphabetically, DB601A, DB601B, et seq.

Jet engines had a rather different classification system. Heinkel had started jet engine experiments in 1936. Their eighth experimental engine became the

HeS8 (indicative of Heinkel, their Stuttgart works formerly belonging to Hirth Motoren which they took over in 1939, and the eighth model). These were works designations, using the prefix approved by the German Air Ministry, in a basic numerical allocation. At this stage the Air Ministry gave the allocations to all jet and rocket engines. In their system the number 019 identified this type of engine and was prefixed by a hyphen to the three figures allotted to individual engine types. The HeS8, as the first, became the HeS 019-001. However, since all these engines were 019, this figure was dropped for general recording and thus the famous Jumo 019-004 is known just as the Jumo 004. However, this point about the 019 prefix is made since it appears in some reference works and will be seen on makers' nameplates. As with internal combustion engines, modifications were notified by suffix letters allotted alphabetically, eg, Jumo 004A. But late in the war came a change which affected the designations for new engines then projected. The last of the three digits was to indicate the manufacturer, by 1 for Heinkel-Hirth, with the prefix HeS indicating the takeover of the Stuttgart works of Hirth by Heinkel. Thus, their first engine was the HeS 011 (with 001 having been allotted, it was the first number in numerical sequence which could end in 1 and so indicate the firm; their next project was therefore the HeS 021). Junkers were allotted the number 2 so that their projects bore the designations Jumo 012 and 022. Argus were given the number 4 for their impulse duct units which powered, for example, the V-1 flying-bomb and they had projects designated As024, As034 and As044. BMW were allotted 8 and their projects under the system became BMW018 and BMW028.

Rocket engines came into the 019 system, but with secondary designations starting at 501 and thereby the Me163's rocket engine evolved by Walter Horten, after the allocation for several experimental units, became the HWK509.

Japan

Pre-war the Japanese manufactured American and European aero engines under licence. In the late '30s Japanese firms held licence for engine construction as follows: Aichi Watch and Electric Machinery Company—400/450 hp Lorraine; Kawasaki Dockyard—BMW types; Mitsubishi—Armstrong Siddeley, Hispano-Suiza and Junkers; Nakajima—Bristol and Lorraine.

The designations used by the Japanese services up to the end of the Second World War are complicated. At the project stage aero engine types for the Army were numbered in a series prefixed 'Ha' for *Hadsudoki* meaning aero engine. On acceptance a type number would be given based on the last two digits of the Japanese year and/or a model number issued in sequence. The engines were also given names. For example, the Nakajima series of 18-cylinder radial engines were given the type name *Homare* meaning honour, and had blocked model numbers in an 11, 12, 21, 22, 41 series. The same series used by the Navy had, for example, NK9B, NK9H and NK9K designations for Army model numbers 11, 21 and 22 respectively. The significance of the Navy's system was that the initial letter denoted the manufacturer, thus N for Nakajima, the second digit gave the class of engine in a brief D for diesel, E for liquid-cooled and K for air-cooled series; the numeral and letter were type and model designators respectively.

The type number given in service could differ from the original type number at prototype stage, so to provide the link between the two differing type numbers the service type number appeared first and the experimental type number last in brackets.

To standardise on engine designations the Japanese Ministry of Munitions introduced a joint service engine identification system using the Ha prefix to denote the series, but with the initial digits denoting engine characteristics. The Nakajima *Homare* Model 22 known by the Navy as NK9K became the *Homare* (Ha-45)22. The number 45 (which should be read as four and five *not* forty-five) indicated an air-coded multi-cylinder engine with a 130 mm by 150 mm bore/stroke in a series in which first digit numbers 1-5 indicated air-cooled engines of differing cylinder numbers and arrangement; 6-7 indicated liquid-cooled engines up to 12 cylinders, and over 12 respectively; 8 and 9 diesel and special engines respectively. Second digit numbers 0-5 indicated 130/160, 140/130, 150/170, 140/150, 140/160, 130/150 bore/stroke in millimetres respectively with 6-9 reserved for further combinations that were not taken up.

Today Japan again has licence for the manufacture of American and British engines, IHI (Ishikawajima-Harima Heavy Industries) having licence from General Electric and Rolls-Royce and Mitsubishi from Allison and Pratt & Whitney. Kawasaki, returning again to aero engine work, is now chiefly involved with repair, overhaul and component manufacture. The American General Electric J79-GE-17 turbojet produced by IHI has the same basic designation, but with the appropriate manufacturer's code change to J79-IHI-17. The Rolls-Royce Turbomeca Adour 801A turbofan has been given an American style designation in the form TF40-IHI-801A by the Japanese who have IHI building it to power their F-1 fighters and T-2 trainers. By the co-operation of the three manufacturers mentioned, turbofan engines of home design are currently being developed, co-ordinated by the Japanese MITI/NAL (Ministry of International Trade and Industry/National Aerospace Laboratory).

United States

American engines were originally designated in much the same way as British engines, by horsepower and manufacturer's name and then by the firm's name and an allotted name in a theme. Two manufacturers in particular dominated the American aero engine scene—Pratt & Whitney (using names like Wasp and Hornet) and Wright (using names like Whirlwind and Cyclone). Then, in addition to the names, the American civil licensing authority imposed their own designating system before the Second World War.

Each engine type was allocated a significant letter relating to the arrangement of cylinders: L for Line (an upright row), O for Opposed (ie, directly opposite), V for Vee-shaped and R for Radial. These prefix letters were hyphenated to the capacity of the cylinders in cubic inches to the nearest multiple of 5 (thus the numbers always ended in 0 or 5). To give a typical example, the Allison V-1710, which powered the P-38 Lightning, P-39 Airacobra and P-40 Tomahawk/Kittyhawk/Warhawk fighters, first appeared in 1930 as the V-1710-A, and a version projected for airship use became the V-1710-B and the first model for service use the V-1710-C. For progressive model changes, letters were allotted alphabetically and minor modifications were designated by number. The first production version was V-1710-C15, which, to spell it out, was the fifteenth modification to the third (C) model of a 'V' banked cylinder engine of approximately 1,710 cu in capacity. There was a further qualification for engines

An example of a horizontally opposed, or flat, engine of six cylinders, in this case an American-designed Continental piston engine built under licence by Rolls-Royce (Rolls-Royce WHP 20977).

adapted for twin-engined aircraft. The first Allisons for the P-38 were the V-1710-F2R and V-1710-F2L, the 'R' and 'L' relating to right and left engines respectively. The number of cylinders did not come into the designating system although to a limited degree it was reflected in commercial names, eg, the Wasp had nine cylinders, the Twin Wasp had 14 and the Double Wasp 18.

So far the designations given have been those of the American aeronautical industry, of which the V-1710 has been a typical example. The US Army Air Corps (later Force) and the US Navy allotted their own combined model and modification number, calling it a series number. The V-1710-C15 was the V-1710-33 in the USAAC inventory and the -F2R and -F2L were -27 and -29 respectively. The series numbers were allotted from No 1 using odd numbers only, while the US Navy used a similar system, but allotting even numbers only. In some cases this meant that an engine ordered by one Service and then later ordered by the other in identical form, would bear a different series number. On the other hand some engines, for example the Pratt & Whitney Wasp, were ordered to a joint Army/Navy (AN) specification and bore designations in the form R-1340-AN-1 upwards using straightforward modification numbering.

The jet engine was first introduced on both sides of the Atlantic for military aircraft and thereby the American gas turbines were designated in a Service system. However, the Americans avoided using the term 'gas turbine' as gas in Americanese was short for gasoline (petrol) thus giving an erroneous impression. Pure jet engines were numbered from 30, prefixed J- for turbojet or T- for turboprop. Later, with the introduction of jet-engined helicopters, the T- signified either turboprop or turboshaft. A further number series from 30 was introduced later prefixed TF- for turbofan. In these series the US Air Force took up odd numbers and the US Navy even numbers. However, this did not always mean that the engine remained exclusive to one Service.

The manufacturer was signified by letters suffixed in a series that has included: A—Allison; BO—Boeing; G—Garrett—AiResearch; GE—General Electric; L—Lycoming; P—Pratt & Whitney; T—Continental; W—Wright; WE—Westinghouse. A further suffix related to the series of modifications with the Air Force taking up odd numbers from 1 and the Navy even numbers from 2. Thus in the case of J-79-GE-7, this would signify the third Air Force modification (by virtue of the odd numbering 1 stood for the initial version and 3, 5 and 7 for the first, second and third modifications) of General Electric (GE) turbojet (J) Type 79 ordered evidently by the USAF (by virtue of 79 being an odd number).

By the late '50s some jet engines were being produced exclusively for civil aircraft and firms' designations came into use. Pratt & Whitney gave their engines type numbers from 1, prefixed JT for jet turbine or TF for turbofan, and suffixed letters and numbers for marks and series. A range of prefix letters are used by other manufacturers and in the main have significance as follows: A or AE—Aerobatic engine; AL—Avco Lycoming; C—Commercial (as distinct from military engines); D—Direct drive; E—Engine; F—Fan; G—Geared; H—for Helicopters; I—Injected (ie, fuel injection); L—Lycoming or left-hand rotation crankshaft; O—Opposed cylinders; PTC—Pratt & Whitney of Canada; R—Right-hand rotation crankshaft; S—Supercharged; T—Turbine or Turbocharged; V—Vertical crankshaft or mounting. These letters may be used in combinations, eg, TFE for turbofan engine.

USSR

Between the wars Russia used an M-prefixed number series for the various engines built to foreign designs, eg, M-15 (Bristol Jupiter), M-25 (Wright Cyclone), M-85 (Gnôme-Rhône K-14). In 1952 the Soviet Union transferred all piston aero-engine production to Poland and engines now bear dual Russo-Polish designations. The Russian jet engines are numbered according to their design bureau with prefix letters relating to the bureau, eg, Ivchenko AI-20 (the AI for Alexander Ivchenko who was design director until 1968) and Kuznetsov NK-8 (the NK for Nikolai Kuznetsov, the director in charge of a bureau).

Chapter 17

Aeronautical armament

To many air enthusiasts armament is a closed book, for it, too, has a language of its own. The scope is wide and the subject can be divided into two main areas: firstly the guns, bombs, missiles, etc, which are carried by aircraft, and secondly the ground weapons designed to destroy aircraft. The most primitive armament was that carried in the early stages of the First World War when flechettes, aerial darts, were used for attacking troops on the ground. The primary weapon of an aircraft to protect itself against others of its kind, the base ground weapon in both World Wars to destroy aircraft, and a current weapon of ground attack, is the gun; this we examine first.

Guns

Guns are classified by the diameter of their bore, called their calibre (pronounced kaliber), their design firm and their model name or designation. Modifications to a model may be by mark number, usually in Roman numerals. For all practical purposes, calibre is also the diameter of the bullets a gun fires, for they must fit the bore snugly to take the maximum blast when fired, and to be turned by the slightly spiralling ridges, called rifling (hence the word rifle) in the bore which turns the bullet on its own axis to ensure its straight flight through the air. Since the bullet fits so precisely, calibres are to three decimal places of an inch; never are they expressed in vulgar fractions. The standard British small arms calibre of both World Wars was .303 in (Point, Three-O-Three) and the Americans used .300 in, although the latter for brevity was often just given as .3 (Point Three). On the continent calibres have always been expressed in millimetres and metric measure is now standard. The NATO standard rifle is 7.62 mm, the equivalent of .300 in. Calibre sizes vary little for the simple reason that weapons are useless unless there is a vast and widespread reserve of ammunition. Any sudden change in calibre would involved a cost of millions of pounds and years of production to produce sufficient of the new ammunition; this ensures standardisation.

To the layman it may sound strange to say that a bullet and a gun are completely harmless, except for toy weapons like air guns which could be lethal. The point is that bullets do not fire themselves, they need an explosive charge and this is placed in a case—a cartridge—and attached to the bullet, the whole being a 'round' of ammunition. Rounds are of several types. Those with a normal pointed metal bullet at the nose are called ball ammunition, a quaint term from the days when the standard ammunition was a ball—of iron! This distinguishes them in name from others with an extremely hard metal nose for armour piercing (AP). There are also rounds without bullets called blanks, used for field exercises, and rounds without charges (explosive) called drill rounds, used for practising gun manipulation.

The foregoing applied to most types of gun, but the rifle, firing shots singly with each squeeze of the trigger, was limited in practice to 1914-15 in air warfare, so our main concern is with the machine-gun which gives so-called 'bursts of fire' with each squeeze of the trigger, by the exhaust gases or part of the explosion being channelled to work the gun. So there is a stream of bullets which, fired from an aircraft at speed, are subject to slipstream, wind and gravity, and so may well have a curved flight path which the gunner cannot see unless their direction is marked—and this is done by bullets, logically called tracer, which mark their passage by smoke by day and light by night. By this means, an air gunner can 'hose' his fire in a similar way to a fireman directing his visible stream of water. There were also incendiary bullets of phosphorous, introduced originally for attacking balloons and airships, which mark their own passage in a similar way to tracer.

Twin Lewis machine-guns on a Scarff ring mounting on an RE8 with ammunition drums in position (left). A German Parabellum, belt-fed but with the belt coiled on a drum (below) (Imperial War Museum).

Machine-guns have to be 'fed' with ammunition. Air machine-guns are normally belt- or drum-fed with the gun action pulling a belt through or turning a drum round. The prime examples of these two different methods are the Vickers and Lewis guns used by the Allies in the First World War, and by many countries between the wars and even in 1939-45. Both guns were of standard rifle calibre to ensure adequate ammunition. Belts of canvas, which got wet and froze, thus causing stoppages, were replaced by connecting links, known as disintegrating links, which fell away from the gun.

Aircraft guns can be mounted in two ways—fixed or free. The Vickers was a fixed gun, which means it was firmly mounted on the structure and was sighted by the pilot aiming the aircraft. In the later stages of the First World War these guns were placed close to pilot eye level and were synchronised with the engine to fire through the propeller arc without hitting the blades. The Lewis gun was not synchronised and when used as a fixed gun it had to be splayed to fire outside the propeller arc. In general the Lewis was the standard free gun, mounted on a pillar or on a cockpit gun ring to permit a gunner to swivel it in any direction as required. This gun was fed from a drum normally holding 39 rounds, with spare drums carried in the cockpit. The German equivalents in the First World War were their 7.92 mm machine-gun designed at Spandau as their standard synchronised fixed gun, and a similar calibre Parabellum machine-gun belt fed by 100-round coils, as their gunners' free weapon.

To fill belts or drums for use was a matter of the armourers knowing the type of operation envisaged in order to prepare the correct mix—the ratio of tracer to ball, perhaps one in five.

As many readers will know, during the Second World War the eight .303 in (expressed in armament terms as 8 × 303) machine-guns of Spitfires and Hurricanes gave way to a smaller number of 20 mm cannon. In aircraft usage a cannon is not a piece of artillery as our dictionaries imply, but a larger type of machine-gun which can fire shells. A shell differs from a bullet in that it is literally a shell, or case, to carry a charge which will explode on impact. Whereas a bullet can only damage by striking, a shell damages by strike and blast. While for artillery shells the charges are often fed to the gun separately, the aircraft cannon is fed with rounds ready for firing. From the First World War to today there has been standardisation, with few exceptions, on 20, 30 and 40 mm calibres for these guns.

The effectiveness of a gun depends largely on the damage its projectiles inflict, but one measurement of efficiency, dependent on the ballistic qualities of the ammunition, is the muzzle velocity. This is simply the speed at which a bullet or shell leaves the end of the barrel, expressed in feet per second in the past and nowadays given in metres per second. Another important factor is the rate of fire which, strangely, is always quoted as rounds per minute (rpm). Why strange? Because you could not fire a machine-gun continuously for a minute! One reason is you would not have that amount of ammunition. For example the .303 Browning guns of a Hurricane I had 332 rounds per gun (rpg) and the guns fired simultaneously at a 1,100 rpm rate. This meant that the guns could fire for only 18 seconds. Now you know why so many aircraft returned to re-arm. Why not carry more ammunition? Well, as it was, 332 rpg equalled the weight of each gun and performance had to be maintained in order to be able to bring the guns to bear. The shells of cannon armament are much heavier and even fewer can be carried. In any case, continuous firing for one minute would wreck many guns

Belts of .300 (top) and .50 ammunition (above) (Ministry of Defence).

by the heat generated through friction. For this reason some modern guns are multi-barrelled. The 30 mm General Electric Avenger cannon has seven barrels which rotate in turn to fire at a maximum of 4,200 rpm. When mounted in a Fairchild A-10 Thunderbolt a drum of only 1,350 rounds is carried—it makes the point. Other important figures in gun specifications for aircraft are weight and overall length.

For the benefit of readers interested in Second World War aviation, it is necessary to mention German guns separately for Germany had its own designating system, 1933-45. Type numbers were given to each type of gun, prefixed by significant initials, eg, MG for *Maschinengewehr* (machine-gun) and MK for *Maschinenkanone* (machine-cannon). An exception to the ruling was the MG151, designed by Mauser as a 15 mm calibre machine-gun, modified for service use as a 20 mm cannon; the designation MG151 was retained, but the calibre change was qualified in the form MG151/20. Anti-aircraft gun types were prefixed BK for *Bordkanone* (aircraft gun) and at least one type was adapted for mounting in aircraft. Special weapons, such as recoilless guns, were prefixed SG. Gun pods were type numbered, prefixed WB for *Waffen Behalter* (weapon container). The WB81, housing downward firing guns, was also known by the name *Giesskanne*, meaning watering can.

The Russians have had their own state system of gun nomenclature from the '20s when their standard 7.62 mm DP machine-gun was adapted for use on aircraft as their DA gun, supplemented later by the improved PV, all of the same calibre. Their aircraft machine-guns of the Second World War date from 1932 when a Soviet design bureau produced a machine-gun capable of 1,800 rpm, designated the 7.62 mm ShKAS (the Sh after V. Shpitalny, head of the bureau). At the same time Shpitalny, in co-operation with S. Vladimirov, designed a larger-calibre gun with the basic designation AK, but with the full title 12.7 mm ShVAK, in recognition of the designers by the incorporation of their initials. They also produced the standard cannon of the Soviet Air Force in the Second World War (known to the Russians as 'The Great Patriotic War 1941-5') as the 20 mm ShVAK and, like the machine-gun, it could be fitted for fire through a propeller boss as well as fitted to turrets.

A team of engineers in 1939 headed by M. Berezin designed a universal 12.7 gun, hence its designation UB, which could be synchronised to fire through the propeller arc or used free or fixed. The 23 mm cannon armament in use in Russia and Soviet satellite forces today, was first introduced in 1941 with the VYa cannon named after the designers A. Volkov and S. Yartsev. Around the same time a 37 mm cannon was devised and entered service as the NS-37, setting the style for post-war gun classifications of calibre, prefixed by initials signifying the designer. Examples of post-war guns are the B-20 (designed by Berezin to replace the VYa), N-37 (replacing the NS-37), NR-23 (coming into service in 1949) and the NR-30 (introduced in the '50s).

Mountings and turrets

From a simple pillar mounting, gun rings were developed to permit a gunner to swivel his weapons and bring them to bear in almost any direction. An RNAS Warrant Officer, F.W. Scarff, designed the most successful type of gun ring which was in use until the Second World War when turrets, to protect gunners from slipstream and the elements, became general. Obviously a gun position will be placed where it can best offer defence of the aeroplane and where a good area can be covered without parts of the aeroplane structure coming into the line of fire. The area which can be covered by the gun is called its field of fire, whether or not it is an area of sky. A field of fire for an air gunner is expressed in degrees of azimuth, elevation and depression which simply means degrees roundwards, upwards and downwards. A gun able to traverse completely all round will have a 360 degree azimuth; but an aircraft's front or rear gunner will be limited to around 180 degrees to avoid hitting part of the aircraft structure. Elevation, as the name suggests, is the highest angle at which the gun can fire from pointing precisely forward and depression is the maximum angle the gun can be depressed from that same level.

The significant data for a turret included the armament (number, calibre and type of guns), its designed position in the aircraft and field of fire, and the weight of the turret and its cupola (that of the FN82 Mk I in a Lancaster weighed 410¾ lb, but the ammunition feed ducts and 2,560 rounds of ammunition weighed just over twice the weight of the turret, without allowing 180 lb for the gunner). The only significant dimension for a turret was the diameter of its turning ring for fitting in the aircraft.

The British turrets were made by aircraft armament specialists and aircraft manufacturers. Turrets were designated by the owners' initials and a type number or letter. Armstrong Whitworth turrets were numbered from AW1, a typical example being the AW15 of the Airspeed Oxford; Bristol, specialising in dorsal turrets, numbered from B1. Boulton Paul types were lettered A, B, C, etc. Fraser Nash types were numbered from FN1 and went into three figures. Modifications to the types were classified by mark numbers, eg, the FN7 in the Botha was the Mk II; the same turret adapted for the Manchester was the Mk III.

Bombs

Of the various explosives used by the air services the most important are bombs. These are classified both by nominal weight in pounds and by type. Since it is a nominal weight, this classification is in round figures. An apparent exception of the past was the 112 lb RL (Royal Laboratory) bombs—but since 112 lb equals one hundredweight (1 cwt) it was, in a way, a round figure. By mid-1918 the RAF had standardised on 20 lb Cooper, 50, 112, 230, 250, 520 and 550 lb bombs for general use, up to 1,650 lb bombs for HP 0/400s and up to a 3,360 lb bomb was planned for the HP V/1500. At the other end of the scale there were BI (baby incendiary) bombs of which 3½ million were ordered, and practice bombs.

Between the wars there was standardisation to 120, 250, 500 lb bombs and during the Second World War there was progression to heavier bombs and a 2,000-pounder was first dropped from a Beaufort, on May 7 1940.

The large bombs were given nicknames. Britain produced large 4,000 lb cylindrical bombs known as 'Cookies' which could be bolted together to make an 8,000-pounder and, for aircraft with specially modified bomb-bays, three could be bolted end-to-end for a 12,000 lb bomb. This should not be confused with the 12,000 lb 'Tallboy' special deep penetration bombs of which 854 were dropped from Lancasters from the night of June 8/9 1944 onwards. The largest British bomb was the 'Grand Slam' of 22,000 lb of which specially modified Lancasters dropped 41 during operations in 1945.

American bombs were in similar weight classifications to the British ones, but not until the Boeing B-29 Superfortress was introduced were bombs of 4,000 lb and over considered. The largest bomb of all was the American T12 of 43,755 lb originally classified as 42,000 and later as 44,000 lb. Too late for the Second World War, the first was dropped inert on March 5 1948 from a B-29. The greatest bomb-load ever was two T12s dropped inert from a Convair B-36 on January 29 1949, representing a 43-ton bomb load.

On the continent bomb weights were in kilograms and the early bombs of the German Air Service were small ones of 4.5 kg and 20 kg. Up to 1916 most German bombs were pear-shaped, like many Allied bombs, but from that year PuW *(Prufenstalt und Werft der Fliegertruppen*—Testing Section and Workshops of the Flying Service) designed torpedo-shaped bombs in 10, 50, 100, 300, 600 and 1,000 kg nominal weights. As with the Allies, small incendiary bombs were produced. In the Second World War there was a simple progression of 250, 500, 1,000 and 2,000 kg bombs.

So far only weight has been mentioned, but type came into all British bomb classifications. The standard bomb was the high explosive (HE) general purpose (GP) type and a 1,000 lb bomb of this type would be classified as 1,000 lb HE(GP). Other types included AP (armour piercing) and SAP (semi-armour piercing). Bombs intended for blast effect where the ratio of weight of explosive to structure was high, were known as HC (high capacity) bombs and with a low ratio LC (low capacity); but many bombs designed to have some measure of penetration and blast were MC (medium capacity). There were special AS (anti-submarine) bombs of 100, 250, 500 and 600 lb as well as depth-charges (DC) which could be set to explode at various depths. Some special bombs came into the category of pyrotechnics, which the dictionary gives as the art of making fireworks, but in armoury terms embraces flares to illuminate targets which slowly descended by parachute, or signal lights to give the recognition colours. The standard signal was the Very light, fired from a pistol and looking much like a Roman candle. Some pyrotechnics come easily into the bomb category since a smoke sea marker was 250 lb and a target indicator (TI), which had to keep burning brightly throughout a bomb attack, could be up to 1,000 lb. Pathfinder aircraft normally carried a 'bomb load' of TIs. With anti-personnel bombs, for attacking troops, there was a range of small bombs classified as F for Fragmentation. The implication is that they were designed to kill by the disintegration of their cases (and sometimes had extra shrapnel added) rather than by blast.

The Germans used type suffix letters to the weight of their bombs in the manner SC1000 (SC for *Spreng-Cylindrisch*—the German equivalent of a general purpose HE bomb), SD for *Spreng-Dickewand*—an HE bomb with thick casing, and PC for *Panzer Cylindrish* meaning armour piercing. Like Allied bombs, some German ones had nicknames as follows: SC1000 'Hermann' (the Christian name of Goering, head of the Luftwaffe), SD1000 'Eseu' (short for *Entseuchung* meaning mine-clearing), PC1400 'Fritz' (a popular German Christian name) and SC1800 'Satan' of which the implication is self-evident, and a 2,500 kg 'Max' bomb. The single 1,400 kg radio-controlled bomb was prefixed FX. German standard anti-personnel bombs of 1 and 2 kg were known respectively as SD1 and SD2 bombs, the latter being nicknamed 'Butterfly Bombs'.

Modifications to bombs, which might affect their filling, case, detonation and setting devices were classified by the British by mark numbers in Roman figures while the Germans used suffix letters. Post-war bomb nomenclature has been modified. American bombs are now classified by mark numbers which each relate to a different size and type of bomb. In spite of metrication and standardisation, most western countries still classify bomb weights in pounds in the series 250, 500, 750, 1,000 and 2,000 lb, while the French and countries of the East continue to use kilograms. In these days of nuclear weapons, 2,000 lb is now the largest standard HE bomb.

The act of bombing has a terminology of its own. Bombs may be carried in three ways; internally in a bay, externally under wings or fuselage, or even on additional stub wings as on a Lysander, or recessed into the fuselage as was the case with 'Tallboys' and 'Grand Slams' on Lancaster B1 (Special) bombers. Bombs are normally carried out to aircraft on low trolleys to go under the aircraft and are winched up (except light bombs which can be manhandled) to be shackled to the aircraft. The manner in which they are released can normally be selected: to drop all at once on a target is called the same as a salute of guns—a 'salvo', but to drop them all for emergency reasons to lighten load is 'to jettison'. To release the bombs one after the other is to drop a 'stick' of bombs.

Where an aircraft has a bomb-bay within the airframe, there are, of course, doors to enclose the structure and which open to let the bombs fall through the aperture. Most modern aircraft have bomb doors which slide and do not lower open into the slipstream, like some bombers of the Second World War. You could easily have been misled on this point, for one kit manufacturer put dropping doors on the model of a famous bomber where the doors in reality slid sideways and upwards into the structure. With many modern bombers acting in multi-roles, and certainly with all maritime aircraft, the bay for bombs, etc, is more appropriately called a weapons-bay.

Quite apart from the variety of air weapons, there are a number of different kinds of bombs currently in use including tactical nuclear bombs, dibber bombs to break up stretches of concrete such as runways, and retarded bombs in which their fall is literally retarded by parachutes so that in a low-level attack the aircraft will be well clear of the target by the time they hit and explode. Then there are cluster bombs which contain a number of smaller bombs (bombphlets) to be scattered and burst over a wide area.

Bombs are fused, that is set to explode, before dropping, in certain cases this can be done in flight, but during the Second World War it was usual to fuze a few hours prior to operations. For impact, the bombs were actually given a fuze of around 0.025 seconds, that minute fraction of a second after hitting ensured a burst at ground level and not roof level, thus achieving maximum blast damage. Also about ten per cent were set to explode on delayed action of some 6 to 144 hours after dropping, to hamper rescue and repair services. War is a very nasty business. One of its nastier aspects is the use of napalm bombs originating from the Second World War when normal fuel tanks were fitted for firing by grenades and dropped to explode and spread fire over a large area. The name napalm comes for naphtha, a highly inflammable liquid.

Torpedoes

Torpedoes were universally classified by their diameter in inches, except by France in particular which used a metric measure. Aircraft-launched torpedoes were, in the main, adapted from standard naval ones. Their significant measurements, apart from the classifying diameter, are length, water speed, total weight and the weight of the warhead, ie, the weight of explosive at the nose. Changes in the composition of warheads and propulsion units led to continuous modifications and torpedoes were additionally classified by type and mark number. Japan also used type numbers but their high figures, such as Type 97, relate to the year of design according to the Japanese calendar not a normal numerical sequence.

Experiments in launching torpedoes from aircraft were planned from 1912 and made from 1913. The first enemy ship sunk by torpedo was a Turkish transport in the Dardanelles hit by a 14 in RGF Mk X torpedo of 1897 vintage launched from a Short 184. As these floatplanes had difficulty in rising with torpedoes, and the 14 in did not have a sufficiently large warhead to menace a warship, Short 320 seaplanes and Sopwith Cuckoo shipborne landplanes were brought into service to carry the 18 in RGF Mk IX torpedo. In all 607 torpedoes were supplied to the RNAS during the First World War.

The Germans also had an air-launched torpedo of 18 in and in late 1918 experiments were in hand for wire-guided air-launched torpedoes from airships. The diameter of 18 in became fairly standard worldwide and the British 18 in RNTF, designed in 1917, was still in use in the Second World War in its Mk IV form. The RNTF stood for the Royal Naval Torpedo Factory at Greenock.

The standard torpedo of the Fleet Air Arm and RAF during the Second World War was the British 18 in Mk XII and the RAF, after trying the USN Mk XIII, went into the post-war years using the British 18 in Mk XV. Germany used two main torpedoes for air dropping, their F5B and the Italian-built F5W Whitehead type. In addition both Germans and Italians used the special LT350 circling torpedo designed by the latter.

In spite of much experimentation, Britain went to the USA for their torpedo weapon in the '50s and bought the American Mk 43. Arabic figures had by this time replaced the Roman numerals used up to XX. Currently both British and United States maritime aircraft carry US Mk 44 and 46 torpedoes.

Rockets

From 1916 electrically fired Le Prieur (after Lieutenant Y.P.G. Le Prieur,

The sting of a Scorpion. The 2.75 in FFARs in the wingtip pod of a Northrop F-89D Scorpion of USAF (Northrop R21478).

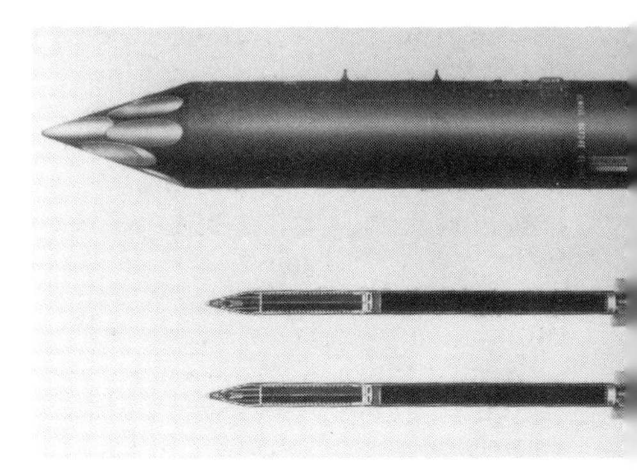

Oerlikon 81 mm folding fin rockets developed in collaboration with SNIA, an Italian organisation, have the name SNORA. These two pictures show SNORA tube launchers suitable for subsonic aircraft such as helicopters and the container or pod required to house the same rockets on high speed fighters (Oerlikon-Bührle).

French Naval Air Service) rockets were fitted to Nieuport Scouts and later to British Sopwith Pups and other types for attacking enemy kite balloons. This form of armament became obsolete by 1918 owing to the introduction of efficient incendiary machine-gun ammunition.

As an alternative to large-calibre guns, rocket armament was re-introduced by the RAF in the Second World War for ground-attack. Rockets were electrically fired from rails fitted beneath aircraft wings and were first used on June 22 1943 on a shipping strike. The rockets were of two operational types, a 60 lb HE 6 in shell and a 25 lb solid AP shell of 3.44 in. Additionally there was a 25 lb practice rocket, originally with a steel, but later with a concrete, head. This type of armament became known throughout the Service as RP for Rocket Projectile.

Now many rockets are podded, being fired in clusters from pods carried beneath the aircraft. They are normally classed by the diameter of their head and are now used in 37, 50, 57, 68, 70, 81, 100, 125 and 135 mm sizes. As it is a once-only firing from their containers, usually known as pods, their diameter is not strictly a calibre for precise fitting, but a housing size.

Abbreviations associated with rockets include FFAR (Fin-folding or Free Flight Aircraft Rocket) and HVAR (High Velocity Aircraft Rocket). Solid fuel unguided aircraft with sliding fins are known as SURA-FL types.

Missiles

When a rocket is guided on its way to the target it comes into the category of a missile, a word that in general usage covers arrows, bullets and even stones. Missiles in the armament world are divided into their appropriate spheres of application as AAM, ASM, AUM, SAM, SSM, SUM, UAM and USM where A is Air, S is surface, U is Underwater and M is missile. Taking it that the word 'to' comes after the first letter, AAM means Air-to-Air Missile, and is thereby a missile carried by aircraft to destroy other aircraft in the air, as opposed to an

SSM which is a Surface-to-Surface Missile. Early SSMs were the German V (*Vergeltungswaffe*—meaning revenge arm) weapons of the Second World War. The V-1 flying-bomb was also a pilotless aircraft, but, by the fact that it carried a warhead, was as much a missile as modern cruise missiles which have the refinement of self-contained guided systems. The V-2 was a giant rocket, but basically similar to the SSMs of today which may now have nuclear warheads and sophisticated guidance systems.

The V-2 was set to go to a certain height and turn towards England with a mechanism to keep it on course; its route was 'pre-set' in missile parlance. But in this age of electronics, information can be given to the missile en route by two main forms of control, guidance and homing.

One form of guidance is beam-riding. The missile is fired into the orbit of a radar beam directed to or tracking the target and it is designed to follow the beam to its target. There is also the so-called command guidance where radars track both target and missile, feed information to a computer and transmit guidance to the missile.

With homing control there are three main forms, passive, active and semi-

Gloster Javelin F(AW)9 XH846 *carrying a Firestreak AAM.*

active. Passive homing implies that the target itself provides the source on to which the missile 'homes', by radiating energy such as light, heat or noise. A heat-seeking missile, for example, would be attracted in particular to vehicles with engines running. An active missile transmits its own source of energy and its target is a source which responds to its transmission. As this means bulky transmission apparatus within the missile, a semi-active missile is used in which a ground or aircraft radar transmits to seek the target and directs the missile accordingly.

Most missiles in service in Western countries have names, examples being Aspide (Italian multi-role), Firestreak (British AAM), Magic (French AAM), Maverick (US AGM), etc, but in addition the United States Services have a standard missile designating system, eg, AIM-7. The number is a type number in a system of mainly significant letters such as: M—Missile; N—Probe; R—Rocket, as the type designating letter, hyphenated to the type number. The letter preceding this (the second letter of the designation) denotes the mission: D—Decoy; E—Electronic (Special); G—Ground (surface attack); I—Intercept; Q—Drone; T—Training; U—Underwater attack; W—Weather. The first letter relates to the launch vehicle or carrier: A—Air; B—Multiple; C—Container; H—Silo-stored; L—Silo-launched; M—Mobile; P—Pad; R—Ship; U—Underwater. Status prefix letters may additionally be placed in front of the designation to denote, as with US aircraft: J—Temporary special test; N—Permanent special test; X—Experimental; Y—Prototype; Z—Planning.

Little is known of the Russian missile type designation, but the NATO designations for Russian missiles are given in the appendices following the NATO names for Russian aircraft. The French words for air and surface are the same as in English so that the prefixes used by Aérospatiale, the main French missile manufacturer, are simply AS for air-to-surface and SS for surface-to-surface missiles.

A number of abbreviations are used with missiles and their systems including: ALCM (Air-Launched Cruise Missile), AMRAAM (Advanced Medium-Range Air-to-Air Missile), EOGB (Electro-Optical Guided Bomb), GW (Guided Weapon), HARM (High-speed Anti-Radiation Missile), HEAT (High Explosive Anti-Tank warhead), Hellfire (Helicopter-launched fire), HOBOS (Homing Bomb System), HOT (High-subsonic, Optically-guided, Tube-launched), LRMTS (Laser-Ranger and Marked-Target Seeker), Milan *(Missile d'Infanterie Léger Anti-char* meaning anti-tank missile system and also known as Euromissile), MIRV (Multiple Independently-targetted Re-entry Vehicles, meaning that the missile contains several warheads which go their separate ways as the target area is neared), SRAM (Short-Range Attack Missile), TOW (Tube-launched, Optically-tracked, Wire-guided).

Atomic and nuclear weapon yields

The power, or yield as it is called, of atomic explosions is measured by relation to tons of conventional high explosive. As such weapons are thousands of times more powerful they are measured in kilotons (thousands of tons) and thermonuclear strategic weapons, or H-bombs, in megatons (millions of tons). It should be appreciated that the references by the Central Office of Information, in giving the damage effects of a 10-megaton bomb with its 20-25 mile radius for area of damage, is a case of taking a round figure and does not necessarily imply that nuclear weapons held by any nation have a yield around that figure.

Chapter 18

The language of the ether

Wireless as a word has given way to radio, but officially in the Services radio was a word which for many years embraced both radar and wireless. The early British Services wireless receivers were called Wireless Sets (WS) and the various types were numbered with WS prefix letters. In the '30s, the RAF classified each main instrument of communications by a number, prefixed by R for Receiver, T for Transmitter or TR for combined Transmitter/Receiver.

Communications are in two main forms, signal and speech. Signal communication, general in the First World War, was by Wireless Telegraphy (W/T) normally using the Morse Code. For those uninitiated in this basic universal code, it is explained that each letter of the alphabet is known by a 'dot' (·) or dash (–) or a combination of both as follows: A ·– B –··· C –·–· D –·· E · F ··–· G ––· H ···· I ·· J ·––– K –·– L ·–·· M –– N –· O ––– P ·––· Q ––·– R ·–· S ··· T – U ··– V ···– W ·–– X –··– Y –·–– Z ––·· 1 ·–––– 2 ··––– 3 ···–– 4 ····– 5 ····· 6 –···· 7 ––··· 8 –––·· 9 ––––· 0 –––––.

On a Morse key sender, the tapper would be touched down and immediately released for a dot (dit), but held down for a fraction of second for a dash (dah). So that an SOS (Save our Souls) emergency signal would have sounded like: dit, dit, dit, dah, dah, dah, dit, dit, dit. The current distress signal for aircraft is to broadcast verbally on the international VHF emergency frequency: "Mayday, Mayday, Mayday, Distress Call. This is . . ." followed by aircraft registration, type, position, height with details of emergency. Such speech is R/T (Radio Telephony).

Speech is, of course, much quicker than signal, where each word is spelt out by letter code. But speech is liable to misinterpretation, especially when there is interference in the ether, so often important words are spelt out by their letters. There, again, confusion can occur with spoken letters, B and P for example sound very much alike. So letters are given as words. This used to be known as the Phonetic Alphabet and was widely used in the First World War and letters were spoken: A—Ac; B—Beer; C—Charlie; D—Don; E—Edward, etc. The Americans used different words, so in the Second World War an Allied Phonetic Alphabet was introduced: A—Able; B—Baker; C—Charlie; D—Dog; E—Easy, etc. Some old-timers (look who's writing!) still use these outdated words. The current correct mode is given in the ICAO Word Spelling Alphabet as follows: A—Alfa; B—Bravo; C—Charlie; D—Delta; E—Echo, F—Foxtrot; G—Golf; H—Hotel; I—India; J—Juliett; K—Kilo; L—Lima; M—Mike; N—November; O—Oscar; P—Papa; Q—Quebec; R—Romeo; S—Sierra;

MAD tails as posed by an RAF Nimrod, the Orions of the RAAF and RNZAF. The MAD tail of the Nimrod has the American designation ASQ-10A and the pod on the fin houses ESM (Electronic Support Measures) equipment.

T—Tango; U—Uniform; V—Victor; W—Whiskey; X—X-Ray; Y—Yankee; Z—Zulu.

Radio equipment is classified according to the frequency band on which it operates; frequency being the wave motion in the number of vibrations a second. Service and civil aeronautics have been mainly concerned with bands MF, HF, VHF and UHF for Medium, High, Very High and Ultra High Frequency respectively.

However, communications are now but one facet of the radiations throughout the ether. Electronic radiations used for detection have led to ECM (Electronic Counter-Measures) and even ECCM (Electronic Counter Counter-Measures) and a whole new range of abbreviations have evolved, including: ASV (Air-to-Surface Vessel radar), AWACS (Airborne Warning and Control System), Chaff (see Window), CRT (Cathode Ray Tube), DME (Distance Measuring Equipment), FLIR (Forward Looking Infra-Red), HUD (Head-up Display—data projected to enable a pilot to have significant readings while looking ahead), IFF (Identification, Friend or Foe), ILS (Instrument Landing System), IR (Infra-Red), Loran (Long-range navigation), MAD (Magnetic Anomaly Detector), Madge (Microwave aircraft digital guidance equipment), Shoran (Short-range navigation), SLAR (Side-Looking Airborne Radar), Tacan (Tactical air navigation), Transponder (equipment triggered by receiving a certain signal).

For the vast range of American equipment there is a relatively simple code to meanings. For example the F-15 Eagle has among its avionics (electronic equipment fitted in aircraft) an APG-63, APX-76 and ALQ-135. You can obtain an idea of the type and function of such equipment by a knowledge of the US military electronic system code.

The first letter relates to where it is installed: A—Airborne; B—Underwater; C—Air Transportable; D—Drone; F—Fixed; G—Ground; M—Mobile; P—Pack; S—Surface craft; T—Transportable (ground); U—Utility; V—Vehicle; W—Water (surface or underwater).

The second letter relates to the type of equipment: A—Infra-red; B—Pigeon; C—Carrier (Wire); F—Photographic; G—Telegraph and Teletype; I—Interphone; J—Electro-mechanical; K—Telemetering; L—Countermeasures; M—Meteorological; N—Noise (sound in air); P—Radar; Q—Sonar; R—Radio; S—Special; T—Telephone (wire); V—Visible light; W—Weapon system; X—Facsimile representation or TV; Y—Data Processing.

The third letter gives the purpose: A—Auxiliary; B—Bombing; C—Communications; D—Direction finding; E—Ejection or release system; G—Fire control; H—Recorder/Reproducer; L—Searchlight control; M—Maintenance or test equipment; N—Navigation aid; P—Reproducing; Q—Special; S—Detecting, ranging or giving bearing; T—Transmitting; W—Controller; X—IFF.

Chapter 19

Airships

Airships (dirigibles) are in a class of their own. Basically an airship is a powered balloon, shaped to reduce resistance to the air. They are divided into two main types, the non-rigid or blimp, which is a shaped gasbag with an underslung 'basket' for power units and personnel, and the rigid type in which gas bags are contained within a shell of skeletal framework covered with fabric.

The efficiency of an airship depends on the load it can carry which is directly related to the capacity of its gasbags. The largest of the First World War airships had a capacity of 2,750,000 cu ft (77,600 cu m). The lengths of the largest were around 700 ft and their top speed rarely exceeded 80 mph. Britain gave up Service airships in 1920 and commercial airships in 1930 after a disaster to the *R101*; Germany discarded airships after their *Hindenburg* crashed in May 1937 at Lakehurst, USA. America used naval airships from the end of the First World War until after the Second and Goodyear have produced a number of small commercial airships which are currently operating.

More airships were in service in 1918 than at any other time and the great majority of all such aerostats which have ever existed were built or designed during the First World War. A few airships were given names, but in general the identity system consisted of an indicative letter or letters and a number as follows:

Belgium

This country was given the Zeppelin *LZ62 (L30)* after the First World War but had it dismantled and broken up in 1920.

Britain

The initial army airships were given names including those of letters of the Greek alphabet in sequence, viz, *Alpha, Beta, Delta,* et seq. When, in 1914, the RNAS took control of all service airships these were re-numbered from No 1 with the prefix 'R' for Rigid. Numbers allocated were consecutive to *R41* and then jumped to *R80* and stopped. The two civil airships ordered in the late '20s were given the numbers *R100* and *R101*.

Blimps were numbered consecutively from No 1 with indicative prefix letters to the numbers allocated. Coastal type C1-27, Improved Coastal type (also known as C Star Type) C*1-10. North Sea type NS1-18, Submarine Scout type SS1-49, Submarine Scout Experimental type SSE1-3, Submarine Scout Patrol type SSP1-6, Submarine Scout Twin-engined type SST1-90, Submarine Scout

Zero type SSZ1-77. Rebuilt blimps had an 'a' suffix after their number and replaced airships a 'b' suffix. A semi-rigid airship acquired from Italy was designated *SR1*.

France

The French Army had airships at the turn of the century, but the French Navy, which created an Airship Service in 1916, took over all military airships in 1917. The French Army named some of their airships, but most French Service airships were numbered consecutively and prefixed by significant index letters: Astra Torres type built by Astra AT1-23, Chalais-Meudon depot built CM1-8, Vendette Twin built by Zodiac V21-23, Zodiac type ZD1-8. Post-war the French received three German Zeppelins: *LZ114* taken into naval service was their *Dixmude, LZ83* (ex-German Army *LZ113*) and *LZ121 Nordstern.* Between the wars the Zodiac Company provided the French Services with Escourteur and Vendette type motor balloons for escort and observation work.

Germany

161 rigid airships were built and flown by the Germans in the period 1897-1940. These were mainly Zeppelins, a product of Luftschiffbau Zeppelin GmbH whose airships (none under 400 ft) were designated a to y according to type and numbered *LZ1-69, 71-114, 120-121* (surrendered to Italy and France respectively), *126* (to US Navy), *127* (the famous *Graf Zeppelin* and the first airship to fly round the world), *129* (the ill-fated *Hindenburg*) and *130* (*Graf Zeppelin II* flown first in 1938 and dismantled in 1940). Early Zeppelins were commercial ventures, but from *LZ3* there were deliveries to the German Army who numbered their first acquisitions *ZI-XII* and then gave the prefix *LZ* to their own non-consecutive numbering—which has caused some confusion to historians. The German Navy took deliveries from *LZ14* (Type h) and these were numbered from *L1,* the *L* for *Luftschiffbau* (airship). These were the airships which, in the main, bombed Britain during 1915-17 and it was by their *L* naval numbers that they were reported, since these were clearly marked on the envelope.

However, not all German airships were Zeppelins, both the German Army and Navy used Schutte-Lanz built airships of types a to f, numbered *SL1-22* by the firm and these numbers were retained as service numbers.

Italy

As early as 1912 the Italian Minister for War had advocated a force of 300 aeroplanes and 24 airships. War in 1915 saw this programme greatly surpassed and by mid-1918 plans were in hand to bring the airship strength to 42. There was, however, only one rigid airship, their G for *Grande* (large) 564 ft in length. Eighteen were semi-rigid M1-18 of which one went to the RAF as their SR1, the M standing for *Medio* (medium). The other types were blimps as follows: A for *Alleggerito* type, DE1-5 *Dirigible Exploratore* (airship experimental) type similar to the British SS, E for *Exploratore* (experimental), F1-6 for *Forlanini* type, P1-9 for *Piccolo* type, SSA for British Submarine Scout erected in Italy plus deliveries, ex-RNAS. U1-10 and V1-20 for *Usuelli* and *Veloce* types respectively. Post-war under the Armistice reparations Italy received two German Zeppelins, *LZ106* (ex-*L61*) named *Italia* and *LZ120* named *Ausonia*.

The Goodyear airship Europa.

Japan

Several non-rigid airships and blimps have been built in Japan but their only large rigid was the ex-Zeppelin *LZ75 (L37)* of which only parts reached Japan.

Russia

The Russian home-built airships were operated by the Army, and for the Navy four coastal blimps, *Ca* to *Cd*, were supplied by Britain. In the '30s the Russians, noting the success of the German *Graf Zeppelin*, started an ambitious airship building programme including types numbered with the prefixes DP and V, but the programme appears to have been abandoned in the Second World War.

United States

The first non-rigid airship for the US Navy was the DN-1 (DN for dirigible non-rigid). Retrospectively this was presumed to have been A1, for all subsequent non-rigids were given type letters from B onwards and were numbered from 1 for each class. At the end of the Second World War there were 8 G-type for training, 134 R-type, 22 L-type and 4 new M-type in service. During the Second World War they made 55,900 patrol sorties escorting 89,000 ship movements. After the war they have continued in anti-submarine warfare patrolling and have been fitted with radar for early warning duties. During the Second World War a new designating system was evolved with subsequent modifications, bringing airship designations in line with those of aeroplanes. For example in the designation *ZP2K-35*, the *Z* would denote LTA (lighter-than-air classification) and was constant for airships of all classes, *P* stood for the function and was selected from a series comprising *H* for search or rescue, *N* (later *T*) for trainer, *P* for patrol, *S* for scout and *U* for utility. The *2* denoted the second-modification and the final letter the type, with the number *35* being the thirty-fifth built of the particular type. For experimental types an '*X*' preceded the whole designation.

Rigid airships were numbered *1* to *5* with pertinent prefixes starting with *ZR* (*Z* to denote LTA and *R* for rigid, as with British airships). *ZR-1* completed in 1923 and named *Shenandoah*, was wrecked in 1925. *ZR-2* was to have been the British *R38* which crashed in the Humber. Two Zeppelins allotted as war reparations were sabotaged and a replacement Zeppelin, the *LZ126*, was built and became the US Navy *ZR-3* and was named *Los Angeles*. Two large rigids designated *ZRS-4* and *ZRS-5* (to denote LTA rigid scouts *4* and *5*) were named *Akron* and *Macon* and commissioned 1931 to 1935. With their wrecking, respectively in 1933 and 1935, the building of rigids was abandoned.

An anomaly in the designating system occurs with the single *ZMC-2* (LTA metal-clad of 200,000 cu ft capacity) used in the period 1929-39.

Appendices

I Works of reference

The prime reference book of British Aerospace is *Jane's All the World's Aircraft*, which has been produced annually for over 70 years, and is sub-titled *The annual record of aviation development and progress*. The sub-title does not do justice to the scope, which goes outside aviation with airship, balloon and space flight development. But each annual volume (the one for 1981 costs £40, and runs to over 800 large format pages not counting advertisement pages) is not large enought to detail all aircraft currently flying. Most aircraft types out of production, have been featured in earlier volumes.

Because Jane's yearbooks are beyond the pocket of most enthusiasts the publishers have introduced a *Pocket Book* series to cover a wide range of interests and available in two price brackets according to whether limp or stiff covers are used. Aircraft subjects in the series are: *Major Combat Aircraft, Commercial Transport Aircraft, Military Transport and Training Aircraft, Light Aircraft, Research and Experimental Aircraft* and *Helicopters*. Additionally, their *World Aircraft Recognition Handbook* is perhaps the most comprehensive book in its field for spotters. For the spotter there is also the Warne's *Observer's Book of Aircraft*, but as a modestly priced pocket book, it is more of an annual addition to the range of aircraft flying, than an annual edition.

Covering similar ground to the Jane's series is the Ian Allan *Of the World* range which includes separate books on Air Forces, Military Aircraft, Civil Aircraft and Helicopters. Certain air arms including, so far, Belgian, Dutch, French, German (Federal) and Spanish and Portuguese (these two countries in the one volume) are covered in a most detailed way by Midland Counties Publications.

For general aviation, the histories of aircraft firms and the aircraft they produced, the Putnam reference series cannot be rivalled and reference is made to these in the chapter on aircraft nomenclature. Aircraft single type histories abound and are too numerous to detail, many cover the same ground as others before them. Due to variations in prices, coverage and standard of reproduction it would be difficult to make judgements. In certain fields, however, there are books which comprehensively cover a subject better than any others produced up to the time of writing. Here is a brief list. *Aircraft of the Royal Air Force*, by Owen Thetford (Putnam) of which, to be up-to-date, the latest edition should be bought. *The Squadrons of the Royal Air Force* by James J. Halley (Air Britain distributed by Midland Counties Publications), gives a brief on all squadrons and includes their RFC or RNAS origins where appropriate. More detail on certain squadrons is given in *Bomber Squadrons of the RAF and their Aircraft* by Philip Moyes and a similar book of Fighter Squadrons by John Rawlings, both published by Macdonald. However, the 'and their Aircraft' sub-title refers to aircraft types used by the squadrons concerned, not specifications of the aircraft as in the book by Owen Thetford. Books on aircraft codes and serials have been dealt with elsewhere in this book.

For comprehensive books on British aviation in the World Wars there are officially sponsored publications. The most detailed account of First World War aeronautics is *War in the Air* by Sir Walter Raleigh (Vol I) and H.A. Jones (Vols II-VI plus III and V map volumes and Appendices volume). Such a set, out of print for many years with an abandoned reprinting in recent times, would run into several hundreds of pounds and is normally only available to the public at certain reference libraries. For the Second World War the three volumes *Royal Air Force 1939-45* by Denis Richards and Hilary St George Saunders, first published by HMSO in 1954, have more recently been re-issued in paperback form. Apart from these three volumes, aviation aspects are dealt with in the many official campaign histories of which *Defence of the United Kingdom* has particular relevance to aerial activities. For the Battle of Britain there are a dozen books, varying in price and presentation. However, for the US 8th Air Force, stationed in the UK, few could contest that Roger Freeman's *The Mighty Eighth* (Macdonald) covers the subject adequately.

Those interested in a book on old aircraft still existing, complete with a guide to their location world-wide, will find *Veteran and Vintage Aircraft* compiled by Leslie Hunt (Garnstone Press) the ultimate guide. For modellers of certain well-known aircraft, who wish guidance in modelling, as well as a history of the aircraft concerned, there is the *Classic Aircraft Series* covering Spitfire, Messerschmitt Bf109, Hurricane, Ju87 Stuka, Lancaster and Mosquito (Patrick Stephens).

Other books have been recommended throughout the text where appropriate. Prices in general have not been given as with new editions prices may change.

II Aeronautical periodicals

Many of us have had our enthusiasm fired by the aeronautical Press and past issues of journals are useful as a comprehensive chronicle of events. Few will hold complete ranges of the past issues because of the costs of binding and the difficulties of storage in domestic accommodation, but they can be perused in some of the central reference libraries of major towns and in specialist libraries. They are our guide to events over the years and the true researcher will check in several periodicals to get more than one angle on the reports.

A representative selection of the more accessible journals is given below. There have been many others; there was an *Aerial Observer* which appeared 1910-11 and *The Aero* ran to seven volumes in 1909-13, but few copies now exist outside the British Museum. These are but two of the aeronautical journals which failed to survive and have become forgotten except by archivists.

Most journals have a number as well as a date and are divided up into volumes. In general, monthly journals have annual volumes and weekly journals have half-yearly volumes. Some journals number issues consecutively irrespective of volume, while others start again at No 1 for each volume. Some go back to No 1 for changes of title, while others continue their numbering system in spite of any change in title. There are no set rules. As a guide to the periods covered and changes in title, a brief is given on the major periodicals over the years. These are presented in alphabetical order of main title.

The Aeroplane appeared weekly from June 8 1911 with 24 pages for 1d. The largest single volume in 1919 ran to 2,572 pages. From April 24 1959 (Issue 2,486, Vol 96) the title changed to *The Aeroplane and Astronautics* and from March 16 1962 (Issue 2,630, Vol 103) to *The Aeroplane and Commercial Aviation News*. The last issue was October 16 1968 (Issue 2,974, Vol 116) on merger with *Flight International*.

Aeroplane Monthly started with the May 1973 issue, published in association with *Flight International* and continues at the time of writing under the editorship of Richard T. Riding. It contains historical articles and modern aircraft are covered in a less technical way than in *Flight International*.

Air Enthusiast, an aeronautical monthly magazine, started in June 1971 under the managing editorship of William Green and editorship of Gordon Swanborough and continues under their direction, but with changes of title to *Air Enthusiast International* from January 1974 (Vol 6, No 1) and to *Air International* from July 1974 (Vol 7, No 1). Both historical and current articles appear on aircraft and air forces, in greater depth than most of the other monthlies, and the magazine is noted in particular for its detailed cutaway drawings. *Air Enthusiast Quarterly* appeared in January 1971 and continues with three issues a year as *Air Enthusiast*.

Air Mail, the Official Organ of the Royal Air Forces Association (RAFA), started as a monthly magazine in May 1944 and by late 1945 it had a certified circulation of 148,000. This series ended in October 1948 (Vol 5, No 6) and it reappeared as a monthly newspaper from New Series, Vol 1, No 1 in November 1948 until 1951 from when it appeared every two months. Unlike the other magazines recorded here this is a magazine not on general sale, being intended for RAFA membership, but it is an important organ to many service aviation enthusiasts as a means of contacting former squadron, station or unit members for research on projects.

Air Pictorial started as the *Air Defence Cadet Gazette* in 1939 but with the change of name of the youth organisation it became the *Air Training Corps Gazette* monthly from March 1941, starting at Vol 1, No 1. From June 1946 (Vol 4, No 5) the name was changed again to *Air Reserve Gazette* with another start at Vol 1 No 1. However, the January 1947 issue started at Vol 9, No 1 with the explanation that it was the ninth volume since the 1939 start and successive volumes continue to this day—but not without further changes of name. From April 1950 the name changed to *Air Pictorial and Air Reserve Gazette* and in the very next issue a larger format was adopted. The final name change to its present *Air Pictorial* came in April 1958 (Vol 20, No 4) when it became the official journal of The Air League. A great advantage with this magazine is that a yearly detachable index is issued without extra charge.

Aircraft Illustrated has appeared monthly since January 1968, launched under the editorship of P.J.R. Moyes and continues today under Martin Horseman. It is published by Ian Allan Ltd who are noted for their ranges of aeronautical and railway books. Additional to the magazine, a series of *Aircraft Illustrated Extras* appeared, replaced later by *Air Extras*, which deal in magazine form with a particular theme.

Aviation News, Britain's only aviation fortnightly, started in November 1972 as an aviation enthusiasts' paper and is now described as Britain's International Aviation Newspaper. It gives a briefing on current events and includes historical articles and has a particular appeal to air enthusiasts largely through the individuality of the proprietor (Alan W. Hall) who is a dedicated enthusiast as well as a professional editor.

Flight appeared weekly from Vol 1, No 1 (New Series) on January 2 1909 as the Official Organ of the Aero Club of the United Kingdom, devoted to 'the interests, practice and progress of aerial locomotion and transport'. Starting as an off-shoot from *The Auto Motor Journal*, it claimed to be the first aeronautical weekly in the world. From January 4 1962 (Vol 81, No 2,756) the title was changed to *Flight International* which continues today.

The Royal Air Force Journal started on November 1 1941 as a 36-page fortnightly to give all RAF ranks Service information not normally released to the Press. Hitherto Service information, apart from orders, had been an RAF weekly bulletin of one or two duplicated sheets, but properly printed from March 1941. The Journal continued monthly until May 1946 (Vol 4, No 5). Each issue had general Service matters except January to March 'War Special Editions' dealing with SEAC (South-East Asia Command), Cranwell and Halton respectively. In its place *Royal Air Force Review* was issued monthly by the Director of Personnel Services, Air Ministry, as a 24-page RAF Journal. After a joint March/April 1947 issue as Vol 1, No 8, it was first put on general sale, price 6d, starting with the August 1947 issue as Vol 2, No 1. From March 1953 (Vol 8, No 6) the title changed to *Royal Air Force Flying Review* and to *Flying Review International* from Vol 19, No 1 in 1962. At this time it claimed to be Europe's biggest selling aviation magazine. From September 1968, the start of Vol 25, the page size was increased and better quality paper used, but the magazine ended with the September 1970 issue (Vol 26, No 9).

The Royal Air Force Quarterly and Empire Air Forces Journal started in January 1930 and ran to over 200 pages per issue. As well as articles on Service matters, there were good briefings on Commonwealth air matters. The size and scope of the Quarterly was considerably reduced during the Second World War and it ended with Vol XX, No 2 in April 1949. A revival as *The Royal Air Force Quarterly and Commonwealth Air Forces Journal* from July 1949 ended with Vol 5, No 3 in July 1953. With a change of title to *Air Power—The Air Forces Quarterly* from October 1953 a start was made again at Vol 1, No 1 and the final issue was in Autumn 1960 (Vol 8, No 1). *The Royal Air Force Quarterly (incorporating Air Power)* again started at Vol 1 in February 1961 with 86-page issues. From Vol 16, the spring 1976 issue, there was an increase of pages with a smaller format and this quarterly finally ended at Vol 18, No 4 with the winter 1978 issue. To replace this within the Service, a magazine *Air Power* was inserted in *Air Clues*, the official monthly professional magazine of the Royal Air Force, which has been running since 1946 and has been restricted to Service personnel.

There are a large number of foreign aviation magazines, particularly American, but few are held in libraries in Britain. For the enthusiast undoubtedly the best is the *Journal of the American Aviation Historical Society* (AAHS) starting in the spring of 1956 with four issues per year. A special index was compiled for the first ten years and thereafter each annual volume has been indexed.

III Aircraft plans

Aircraft plans to an enthusiast are a loose term for aircraft general arrangement drawings. The actual drawings, or plans, for any one production aircraft type run into thousands and to acquire copies of these drawings from the designing firm would cost a small fortune, apart from the accommodation which you would need to house them. For record purposes most firms now use microfilm for storage of drawings, and the only people outside the industry concerned with acquiring them are the replica builders and restorers.

General arrangement drawings

The aircraft plans to which the enthusiast refers are the general arrangement drawings of an aircraft, sometimes called the three-view drawing because they show the three most significant views of: plan (looking directly down from above); side elevation or profile; head-on (rarely, if ever, the view from the rear).

Plans, even when intended for modellers, should be sound engineering drawings. Such drawings start with a plan view and the first and most important line of all to be drawn will be an imaginary one as far as the aircraft itself is concerned—its datum line running down the centre of the fuselage. This line is shown broken, or should be, to indicate that it is not part of the aircraft's structure. All features are drawn from this datum line, eg, the tip of one wing would be marked off from this line on dividers which would then be swung round to mark off the opposite wing tip.

The profile view, called the side elevation, is projected from the plan view. The draughtsman does not re-measure the fuselage length for this view, or where the wing or tailplane will come. He has done that for the plan. He draws another datum below his plan with his T-square and using a set square on this drops vertical lines down to give him these measurement points. These working lines are rubbed out afterwards. The professional uses the right side of the right sort of paper and pencils of varying hardnesses. His head-on view is similarly projected at the side of the plan view. However, when it comes to presentation in a book the drawings are often in the wrong format to fit in and so are re-arranged one under the other to fit the page shape and size.

Three-view drawings will vary in some respects according to the purpose for which they are produced. Plan view implies the view from above, but three-view drawings for aircraft recognition purposes show the plan view from below. The plan form is the same, of course, for either view, but detail such as engine nacelles will differ. The head-on view may show the aircraft at rest, but for aircraft recognition the flying view with undercarriage retracted, where appropriate, should always be shown. Often for recognition purposes the drawings are shown in black with detail lines in white—in this case it is known as a three-view silhouette. In the attractive Blandford colour series, where a plan and side elevation are given in colour, the plan view is split to show one half from above and one half from below, which makes the plan appear strange, but is quite acceptable as most aircraft structures are symmetrical and it is a way of including additional information. Such presentation has a shortcoming, however, in that there is no clear indication of asymmetrical markings such as the port-upper-wing and starboard-lower-wing-only presentation of the United States insignia on aircraft of USAF and USN.

Scale

General arrangement drawings are presented in a number of different scales according to the purpose for which they are required. The standard scale was 1:72 for pre-war balsa-wood modelling and for most post-war plastic kits. At this scale, a 72nd of actual size, 1 in equals 6 ft, so that if you had an aircraft of 30 ft span, the drawing would span 5 in. A large aircraft, like a Boeing 747 with a span approaching, and length over 200 ft, cannot be presented in a normal book at standard scale, so a 1:144, or even smaller scales, are usual, in the same way that plastic model kits for the larger aircraft are often marketed in smaller scales. The French have gone along with a 1:72 scale, but with metric measures 1:100 and 1:200 scalings would ease calculations for those producing drawings and it is significant that some model armoured fighting vehicle manufacturers are using these scales. At the other end of the market, some skilled modellers prefer larger scales such as 1:48 to permit the inclusion of greater detail on models.

Plan services

For the specialist modeller, plan and data collector, there are plan services. Modellers have for years been served by the Model & Allied Publications Plans Service and they produce a catalogue titled *Scale Drawings—Plans Handbook*. The handbook, available from model shops, details plan scales and prices and advises on additional information given, such as cockpit detail.

An Aircraft Plans Service is also offered by the enterprising aeronautical fortnightly *Aviation News* which, being of newspaper format, has plans on its centre pages. The range of plans available appears at frequent intervals in *Aviation News*.

IV **Civil aircraft registration markings** (see chapter 7)

Several reference books give the current registrations. Here current and post registrations are given for the benefit both of aircraft spotters and researchers of records and photographs. Non-current registrations are shown in italics.

* Denotes that numbers may be used after this prefix instead of letters.

A-	*Austria* (to 1939)	CU-	Cuba
AN-	Nicaragua	CV-	Romania
AP-	Pakistan	CX-	Uruguay
A2-	Botswana	CY-	*Ceylon*
A6-	United Arab Emirates	C2-	Nauru (Pacific)
B-	Formosa (Taiwan)	D-	*Germany* (1919-45)
CC-	Chile	D-	Federal Germany
CCCP-	Soviet Union	DM-	East Germany
CF-	Canada	DQ-F	Fiji
CF-Z	Canadian gliders	D6-	Comoro
CH-	*Switzerland*	EC-	Spain
CN-	Morocco	EI-	Eire
CP-	Bolivia	EL-	Liberia
CR-	Portugal (overseas)	EP-	Iran
CR-C	Cape Verde Republic	ET-*	Ethiopia
CS-	Portugal	EZ-	*Saarland* (pre-1935)

F-	France
FC-	*Free French* (1940-5)
FD-	*French Morocco*
F-KH	*Cambodia*
F-L	*Laos*
F-O	French overseas
F-OG	Guadeloupe
F-VN	*Vietnam*
F-W	French prototypes
G-	Great Britain
G-AU	*Australia*
G-C	*Canada*
G-E	*British Isles*
G-F	British airships
G-I	*India* (to 1929)
G-K	*Kenya*
G-NZ	*New Zealand*
G-UA	*South Africa*
HA-	Hungary
HB-	Switzerland and Liechtenstein
HC-	Ecuador
HH-	*Colombia*
HH-	Haiti
HI-*	Dominica
HJ-	*Colombia*
HK-*	Colombia
HL-	Republic of Korea
H-N	*Holland* (pre-1934)
HP-*	Panama
HR-	Honduras
HS-*	Thailand
HZ-	Saudi Arabia
L-	Italy
J-	*Japan* (pre-war)
JA-*	Japan
JY-	Jordan
J2-	Djibouti
K-S	*Finland*
LG-	*Guatemala*
LN-	Norway
LQ-	Argentina
LR-	*Lebanon* (1945-51)
LV-	Argentina
LX-	Luxembourg
LZ-	Bulgaria
MC-	*Monte Carlo*
N-*	United States of America
NC-*	*USA commercial*
NL-*	*USA limited licence*
NM-	Cuba
NP-	*Philippines*
NR-*	*USA restricted flying*
OA-	*Peru*
O-B	*Belgium* (to 1929)
OB-	Peru
OD-	Lebanon
OE-	Austria (post-1955)
OH-	Finland
OK-	Czechoslovakia
OO-	Belgium (since 1929)
OY-	Denmark (since 1929)
P-B	*Brazil* (to 1929)
P-P	*Poland* (to 1930)
PH-	Netherlands
PI-*	Philippines
PJ-	*Curacao*
PJ-	Netherlands Antilles
PK-	*Dutch East Indies*
PK-	Indonesia
PP-	Brazil
PT-	Brazil
PZ-	Surinam
RI-*	*Indonesia*
RV-	*Persia*
RX-*	*Panama*
RY-	*Lithuania*
S-	*Sweden* (pre-1930)
S-P*	*Panama*
SA-	*Saudi Arabia*
SE-	Sweden
SL-	*Saarland*
SN-	*Sudan*
SP-	Poland
SR-	*Syria* (to 1951)
ST-	Sudan
SU-	Egypt
SX-	Greece
S2-	Bangladesh
T-	Denmark (to 1929)
TC-	Turkey
TF-	Iceland
TG-	Guatemala
TI-*	Costa Rica
TJ-	*Transjordan*
TJ-	Cameroun
TL-	Central African Republic
TN-	Congo (Brazzaville)
TR-	Gabon
TS-	*Saarland* (pre-war)

TS-	Tunisia
TT-	Chad
TU-	Ivory Coast
TY-	Dahomey
TZ-	Mali
UH-	*Hedjaz*
UL-	*Luxembourg*
UN-	*Yugoslavia*
VO-	*Newfoundland*
VP-A	*Gold Coast*
VP-B	Bahamas
VP-C	*Ceylon*
VP-F	Falkland Islands
VP-G	*British Guiana*
VP-H	Belize (British Honduras)
VP-J	*Jamaica*
VP-K	*Kenya*
VP-L	Antigua
VP-M	*Malta*
VP-N	*Nyasaland*
VP-P	Western Pacific Isles
VP-R	*Northern Rhodesia*
VP-S	*Somaliland*
VP-T	*Trinidad & Tobago*
VP-U	*Uganda*
VP-V	St Vincent
VP-W	*Rhodesia*
VP-W	Zimbabwe
VP-X	Gambia
VP-Y	*Rhodesia*
VP-Y	Zimbabwe
VP-Z	*Zanzibar*
VQ-C	*Cyprus*
VQ-F	Fiji
VQ-G	Grenada
VQ-L	St Lucia
VQ-P	*Palestine*
VQ-Z	*Basutoland, Bechuanaland and Swaziland*
VR-A	*Aden*
VR-B	Bermuda
VR-G	Gibraltar
VR-H	Hong Kong
VR-J	*Johore*
VR-J	Belize
VR-L	*Sierra Leone*
VR-L	Cayman Islands
VR-N	*Nigeria*
VR-O	*North Borneo*
VR-R	*Federated Malay States*
VR-S	*Singapore*
VR-T	*Tanganyika*
VR-U	Brunei
VR-W	*Sarawak*
VR-	India
X-A/B	*Mexico*
XA-	Mexico commercial
XB-	Mexico private
XC-	Mexico state-owned
XH-	*Honduras*
XT-	*China*
XT-	Upper Volta
XU-	Kampuchea (ex-Cambodia)
XV-	Vietnam
XW-	Laos
XY-	Burma
YA-	Afghanistan
YE-	*Yemen*
YI-	Iraq
YJ-	*New Hebrides*
YK-	Syria
YL-	*Latvia* (to 1940)
YM-	*Danzig*
YN-	*Nicaragua*
YR-	Romania
YS-*	El Salvador
YU-	Yugoslavia
YV-	Venezuela
ZA-	Albania
ZK-	New Zealand
ZP-	Paraguay
ZS-	South Africa
3A-	Monaco
3C-	Equatorial Guinea
3D-	Swaziland
3X-	Guinea
4R-	Sri Lanka
4W-	Yemen (North)
4X-	Israel
5A-	Libya
5B-	Cyprus
5H-	Tanzania
5N-	Nigeria
5R-	Malagasy
5T-	Mauretania
5U	Niger
5V-	Togo
5W-	Western Samoa

A typical civil registration presentation, shown on a Swiss-registered Praga E114M Air Baby, constructor's number 119, previously French-owned as F-BCSN.

5X-	Uganda	8R-	Guyana
5Y-	Kenya	9G-	Ghana
60S-	Somalia	9H-	Malta
6V-	Senegal	9J-	Zambia
6W-	Senegal	9K-	Kuwait
6Y-	Jamaica	9L-	Sierra Leone
7O-	South Yemen	9M-	Malaysia
7P-	Lesotho	9N-	Nepal
7Q-	Malawi	9Q-	Zaire
7T-	Algeria	9U-	Burundi
7Y-	Benin	9V-	Singapore
8P-	Barbados	9XR-	Ruanda
8Q-	Maldives	9Y-	Trinidad and Tobago

V **British Class 'B' (provisional) registrations** (see chapter 7)

Where a letter was used prior to 1945, this is given in brackets. Parnall (J), Martin-Baker (MB) and Weir (W) ceased manufacture before 1945. Letter Q was not allotted.

G-1-	AWA (A-), later HSA, later R-R (1971) Ltd	G-3-	Boulton Paul (C-), but not used
G-2-	Blackburn (B), then HSA (Brough)	G-4-	Portsmouth Aviation (D-), later Miles Aviation

G-5- De Havilland (E-), later HSA (Hatfield and Chester)
G-6- Fairey (F-), later Westland (Fairey Division)
G-7- Gloster (G-), later Slingsby
G-8- Handley Page (H)
G-9- Hawker (I), later HSA
G-10- Reid and Sigrist
G-11- Avro (K-), later HSA (Woodford)
G-12- Saro (L-), later Westland
G-13- (not allotted)
G-14- Short Bros (M-) later Short and Harland
G-15- Vickers (Supermarine) (N-), later Vickers Armstrong (Engineers)
G-16- Vickers (O-), later BAC
G-17- Westland (P-)
G-18- Bristol (R-), later BAC
G-19- Heston Aircraft (S-)
G-20- General Aircraft (T-)
G-21- Miles (U), later HP (Reading)
G-22- Airspeed (V-)
G-23- Percival (X-), later BAC (Luton)
G-24- Cunliffe Owen (Y-)
G-25- Auster (Z-)
G-26- Slingsby
G-27- English Electric, later BAC
G-28- BEA, later BEA (Helicopters)
G-29- Napier
G-30- Pest Control Ltd
C-31- Scottish Aviation
C-32- Cierva
G-33- Flight Refuelling Ltd
C-34- Christlea
G-35- F.G. Miles, later Beagle
G-36- College of Aeronautics and Institute of Technology
G-37- Rolls-Royce
G-38- DH (Propellers), later Hawker Siddeley Dynamics
G-39- Folland, later BSA (Hamble)
G-40- Wiltshire School of Flying
G-41- Aviation Traders (Engineering)
G-42- Armstrong Siddeley
G-43- Edgar Percival
G-44- Agricultural Aviation
G-45- Bristol-Siddeley Engines
G-46- Westland Helicopters
G-47- Lancashire Aircraft
G-48- Westland (Weston Division)
G-49- F.G. Miles Engineering
G-50- Alvis Ltd, Coventry
G-51- Britten Norman
G-52- Marshall's Engineering

The initial allocation to Percival aircraft post-war; their No 1 shown was a Percival T16/45 prototype, photographed in January 1951.

VI NATO code-names for Soviet aircraft and missiles

Aircraft

Since the Soviet Union does not readily distinguish its aircraft by designation, a code-name is adopted. This is particularly important in aircraft recognition reporting where training aims at making a shape evoke a name. The need for this is evident when it is realised that the Press originally reported the MiG-23 as Foxbat, whereas this twin-tailed fighter was actually the MiG-25.

The coding is by a single letter with an indicative initial letter: B—Bomber; C—Cargo/Transport/Airliner; F—Fighter; H—Helicopter; M—Miscellaneous including trainers. Single-syllable words indicate a propeller-driven aircraft, eg, Bear, and twin syllables a jet-engined aircraft, eg, Badger.

Only the abbreviated designation is given as the system of type naming is given in chapter 5. * denotes aircraft now considered obsolete.

Code-name	Aircraft
Backfin	Tu98*
Backfire	Tu-22M
Badger	Tu-16
Bank	B-25 Mitchell*
Barge	Tu-85*
Bark	Il-2*
Bat	Tu-2*
Beagle	Il-28
Bear	Tu-20/95
Beast	Il-10*
Beauty	Renamed Blinder
Bison	Mya-4
Blinder	Tu-22
Blowlamp	Il-54*
Bob	Il-4*
Boot	Tu-91*
Bosun	Tu-14*
Bounder	Mya-52*
Box	A-20 Boston/Havoc*
Brassard	Renamed Brewer
Brawny	Il-40*
Brewer	Yak-28
Buck	Pe-2*
Bull	Tu-4*
Butcher	Tu-82*
Cab	Li-2/DC-3 Dakota
Camber	Il-86
Camel	Tu-104
Camp	An-8
Candid	Il-76
Careless	Tu-154
Cart	Tu-70*
Cash	An-28
Cat	An-10*
Charger	Tu-144
Clam	Il-18* (original)
Clank	An-30
Classic	Il-62
Cleat	Tu-114
Cline	An-32
Clod	An-14
Coach	Il-12
Coaler	An-72
Cock	An-22
Coke	An-24
Colt	An-2
Cooker	Tu-110*
Cookpot	Tu-124
Coot	Il-18 (current)
Cork	Yak-16*
Crate	Il-14
Creek	Yak-12* (later)
Crib	Yak-8*
Crow	Yak-12* (original)
Crusty	Tu-134
Cub	An-12
Cuff	Be-30
Curl	An-26
Faceplate	cancelled
Fagot	MiG-15
Faithless	STOL*
Fang	La-11*
Fantail	La-15*
Fargo	MiG-9*
Farmer	MiG-19
Fearless	cancelled
Feather	Yak-17*
Fencer	Su-19
Fiddler	Yak-28
Fin	La-7*

Firebar	Yak-28P
Fishbed	MiG-21
Fishpot	Su-9
Fishpot C	Su-11
Fitter A	Su-7
Fitter B	Su-17
Fitter C	Su-20
Fitter E	Su-22
Flagon	Su-15
Flashlight	Yak-25
Flipper	cancelled
Flogger B/C/E	MiG-23
Flogger D	MiG-27
Flora	Yak-23*
Forger	Yak-36
Foxbat	MiG-25
Frank	Yak-9*
Fred	P-63 Kingcobra*
Freehand	Yak-?*
Fresco	MiG-17
Fritz	La-9*
Halo	Mi-?
Hare	Mi-1
Harke	Mi-10
Harp	Ka-25*
Hat	Ka-10*
Haze	Mi-14
Hen	Ka-15*
Hind	Mi-24
Hip	Mi-8
Hog	Ka-18*
Homer	Mi-12
Hoodlum	Ka-26
Hook	Mi-6
Hoop	Ka-22*
Hoplite	Mi-2
Hormone	Ka-25K
Horse	Yak-24*
Hound	Mi-4
Madge	Be-6
Maestro	Yak-28UT1
Magnet	Yak-17UT1
Magnum	Yak-30
Maiden	Su-9U
Mail	Be-12
Mallow	Be-10*
Mandrake	Yak-?*
Mangrove	Yak-28
Mantis	Yak-23
Mare	Yak-14*
Mark	Yak-7V*
Mascot	I1-28UT
Max	Yak-18
May	I1-38
Maya	L-29 Delfin
Midget	MiG-15UT1
Mink	UT-2*
Mist	Ts-25*
Mole	Be-8*
Mongol	MiG-21UT1
Moose	Yak-11*
Mop	GST/Catalina*
Moss	Tu-126
Mote	MBR-2*
Moujik	Su-7UT1
Mug	MDR-6*
Mule	Po-2

Air-to-Air Missiles

(Aircraft type operating given in brackets)

AA-1	Alkali (Farmer)
AA-2	Atoll (Fishbed)
AA-3	Arab (Firebar)
AA-4	(withdrawn ?)
AA-5	Ash (Fiddler)
AA-6	Acrid (Foxbat)
AA-7	Apex (Flogger)
AA-8	Alphid (Flogger)

Air-to-Surface Missiles

AS-1	Kennel (Badger)
AS-2	Kipper (Badger)
AS-3	Kangaroo (Bear)
AS-4	Kitchen (Blinder)
AS-5	Kelt (Badger)
AS-6	Kingfish (Backfire)
AS-7	Kerry (Fencer)

Mobile Surface-to-Air Missile Systems

SA-1	Guild
SA-2	Guideline
SA-3	Goa
SA-4	Ganef
SA-5	Griffin/Gammon
SA-6	Gainful
SA-7	Grail
SA-8	Gecko
SA-9	Gaskin

Surface-to-Surface Missiles

SS-1	Scud
SS-4	Sandal
SS-5	Skean
SS-7	Saddler
SS-8	Sasin
SS-9	Scarp
SS-10	Scrag
SS-11	Sego
SS-12	Scaleboard
SS-13	Savage
SS-14	Scamp
SS-15-20	?

Shipborne (Naval) Missiles

SA-N-1	Goa
SA-N-2	Guideline
SA-N-3	Goblet
SS-N-1	Scrubber
SS-N-2	Styx
SS-N-3	Shaddock
SS-N-4	Sark
SS-N-5	Serb
SS-N-6	Sawfly
SS-N-7-10	?

VII Allied code-names for Japanese aircraft types during the Second World War

It should be appreciated that this is a list of code-names allotted by Allied Intelligence to aircraft types assumed as likely to be operated by Japanese forces; not all listed were used and the Allies were unaware of some other types in limited service or under development.

Abdul	Nakajima Ki-27
Adam	Nakajima SKT-97
Alf	Kawanishi E7K2
Ann	Mitsubishi Ki-30
Babs	Mitsubishi Ki-15 and C5M
Baka	Yokosuka MXY7
Belle	Kawanishi H3K1
Ben	Nagoya Sento KI-001
Bess	Heinkel He111
Betty	Mitsubishi G4M/G6M
Bob	Kawasaki Ki-28
Buzzard	Kokusai Ku-7
Cedar	Tachikawa Ki-17
Cherry	Yokosuka H5Y
Clara	Tachikawa Ki-70
Claude	Mitsubishi A5M4
Cypress	Kyushu K9W
Dave	Nakajima E8N
Dick	Seversky 2PA-B3
Dinah	Mitsubishi Ki-46, Army 100
Doc	Messerschmitt Me110 (sic)
Doris	Mitsubishi B-97
Dot	Judy also allotted to same type
Edna	Mansyu Ki-71
Emily	Kawanishi H8R
Eva	Mitsubishi Ohtari (not a bomber)
Frances	Yokosuka P1Y
Frank	Nakajima Ki-84
Fred	Focke-Wulf FW190A
Gander	Kokusai Ku-8
George	Kawanishi N1K
Glen	Yokosuka E14Y
Goose	Code-name changed to Gander
Grace	Aichi B7A
Gus	Nakajima AT-27
Gwen	Name changed to Sally
Hamp	Name changed to Zeke 32
Hank	Aichi E10A
Hap	Original name given to Hamp
Harry	Mitsubishi TK-4
Helen	Nakajima Ki-49
Hickory	Tachikawa Ki-54
Ida	Tachikawa Ki-36/55
Ione	Aichi AI-104
Irene	Junkers Ju87
Irving	Nakajima J1N1
Jack	Mitsubishi J2M
Jake	Aichi E13A
Jane	Name changed to Sally

Janice	Junkers Ju88
Jean	Yokosuka B4Y
Jerry	Heinkel He112
Jill	Nakajima B6N
Jim	Original allocation for Oscar
Joe	TK-19
Joyce	Allotted in error
Judy	Yokosuka D4Y
Julia	Allotted in error
June	Allotted in error
Kate	Nakajima B5N
Laura	Aichi E11A
Lily	Kawasaki Ki-48
Liz	Nakajima G5N
Lorna	Kyushu Q1W
Loise	Mitsubishi Ki-2
Luke	Mitsubishi J4M
Mabel	Original name for Kate
Mary	Kawasaki Ki-32
Mavis	Kawanishi H6K
Mike	Messerschmitt Me109 (Bf109)
Millie	Vultee V-11GB (believed being built)
Myrt	Nakajima C4N
Nate	Duplicated allocation for Abdul
Nell	Mitsubishi G3M
Nick	Kawasaki Ki-45
Norm	Kawanishi E15K
Norma	Allotted in error
Oak	Kyushu K10W
Omar	Suzukaze 20
Oscar	Nakajima Ki-43
Pat	Tachikawa Ki-74
Patsy	Name changed from Pat
Paul	Aichi E16A
Peggy	Mitsubishi Ki-67
Perry	Kawasaki Ki-10
Pete	Mitsubishi F1M
Pine	Mitsubishi K3M
Randy	Kawasaki Ki-102
Ray	Allotted in error
Rex	Kawanishi N1K
Rita	Nakajima G8N
Rob	Kawasaki Ki-64
Rufe	Nakajima A6M2
Ruth	Fiat BR20
Sally	Mitsubishi Ki-21

Good flying shots of Second World War Japanese aircraft are rare, except for aircraft captured and photographed at leisure, as in the case of this Jill being flown by TAIU (Technical Air Intelligence Unit) SWPA (South-West Pacific Area) in June 1945 (US BuAer 194921).

Sam	Mitsubishi A7M
Sandy	Same type as Claude
Slim	Watanabe E9W
Sonia	Mitsubishi Ki-51
Spruce	Tachikawa Ki-9
Stella	Kokusai Ki-76
Steve	Mitsubishi Ki-73
Susie	Aichi D1A1-2
Tabby	Douglas DC-3
Tess	Douglas DC-2
Thalia	Kawasaki Ki-56
Thelma	Lockheed 14-SS1
Theresa	Kokusai Ki-59
Thora	Nakajima Ki-34
Tillie	Yokosuka H7Y
Tina	Allotted in error
Toby	Lockheed 14
Tojo	Nakajima Ki-44
Tony	Kawasaki Ki-61
Topsy	Mitsubishi Ki-57
Trixie	Junkers Ju52/3M
Trudy	Focke-Wulf FW200
Val	Aichi D3A
Willow	Yokosuka K5Y
Zeke 11	Mitsubishi A6M1
Zeke 21	Mitsubishi A6M2
Zeke 22	Mitsubishi A6M3 modified
Zeke 32	Mitsubishi A6M3
Zeke 52	Mitsubishi A6M5

VIII Calibres of aircraft and anti-aircraft weapons

The calibres are given in ascending order by the calibre by which they were known, eg, 20mm guns were the equivalent of 0.787 in calibre, but were invariably called by their metric measure. Abbreviations: AA for anti-aircraft, HAA for Heavy AA and LAA for Light AA, mg for machine gun and QF for quick-firing.

Millimetres	Inches	Remarks
6.5	—	Italian Revelli mg of WW1*.
7	—	Automatic carbines for German aviation troops WW1.
7.5	—	French fighters of '30s.
7.62	0.300	US mgs of WW1 and 2, and still current for light attack aircraft. Standard rifle calibre past and present.
7.7	0.303	British standard mg and rifle calibre in WW1 and 2, also used by other countries. Included Lewis, Browning, Vickers, etc, mgs.
7.92	—	Standard German aircraft defence mg WW2, also Japanese and Swedish guns.
9	—	Villa Perosa Italian mgs WW1, currently ground use only.
11	—	Vickers anti-balloon gun introduced late 1918 by re-boring 500 .303 guns for special tracer.
—	0.44	4,000 Remington rifles from USA for RNAS, 1915.
—	0.45	500 Martini-Henry rifles from trade for RNAS, 1914.
12.7	0.5	Standard American, Italian, Japanese, Russian, etc, mgs introduced WW1, used throughout WW2, including limited RAF use, and currently in use for ground attack and LAA guns.
13	—	German MG215 as originally planned. Standard German bomber turret mg (MG131) of WW2.
15	—	German Mauser MG151 as first produced.
20	—	Introduced WW1. Standard WW2 aircraft cannon and LAA weapon (Hispano, Oerlikon, Madsen, Polsten, etc), German MG151/20 and MK213 aircraft defence guns and 2 cm Flak guns. Also used by Russians and Japanese and remains a standard calibre.
23	—	Russian cannon guns past and present and currently of Polish aircraft armament.
25	—	French (Hotchkiss), Japanese and Russian LAA guns from the '30s.
27	—	Current Mauser cannon, used by Alpha Jet and Tornado.
30	—	Standard cannon calibre including German MK101, improved MK103 and MK108 of WW2 and current Aden, DEFA, Oerlikon KCA, etc.
35	—	Prototype German airborne recoilless gun, 1945 and Italian Type 102 LAA guns.

* For the sake of brevity in this table WW1/WW2 refer to the First and Second World Wars.

A 40 mm Bofors anti-aircraft gun used during the Second World War and into the '70s, seen here operated by the RAF Regiment.

Millimetres	Inches	Remarks
37	1.457	Limited AA use WW1. Standard LAA guns WW2 by Americans, French (Hotchkiss), Germans (Flak 18, 36, 37, 43 types), Italians and Russians. Limited use airborne strafing gun of WW2. Current standard rocket size.
40	1.575	WW1 and 2 standard LAA guns including Bofors, the most widely used AA gun of WW2. Calibre of special 'tank busting' Rolls and Vickers S guns carried by Hurricanes in WW2 and of air transportable guns. Also calibre of helicopter standard grenade launchers.
—	1.59	Vickers QF guns used on NS airships as anti-submarine weapon.
46	—	Italian mobile AA gun.
50	—	German and Italian AA guns. Converted for air-mounting but abandoned.
53	—	Italian multi-purpose gun used as AA weapon.
55	—	German MK112 (scaled up MK108) planned for 1945.
57	2.244	AA guns of WW1 and limited use aircraft cannon of WW2. Standard aircraft rocket size.
68	—	Standard aircraft rocket size.
70	2.75	Standard aircraft rocket size. Japanese AA barrage mortar of WW2.

Millimetres	Inches	Remarks
75	2.95	Anti-aircraft guns from WW1 in Belgium, Czech, French, German, Italian, Japanese, Norwegian, Polish and Russian service. Limited conversion for air mounting and air transit. Bofors gun of this calibre developed in '20s and currently Bofors rocket size.
76.2	3.00	Standard British, American and Russian AA guns of WW1 still in use in WW2. Experimental 12-pounder gun for airships in WW1. Italian and Japanese AA guns of WW2.
76.5	3.01	Austrian field guns captured WW1 and adapted WW2 by Italy for AA defence.
80	—	Continental AA guns WW2.
81	—	Standard infantry mortar size, used by Japanese for anti-aircraft barrage mortars WW2.
83.5	3.287	Limited use anti-aircraft guns between World Wars.
—	3.3	Limited adaption by British of 3 in guns for AA use WW1.
85	—	Russian AA guns introduced 1939.
88	—	Standard German artillery calibre also widely used by Germans, Italians and Japanese in AA role.
90	—	Standard Allied and Italian AA guns of WW2.
—	3.6	British caterpillar track mounted AA gun, 1918-27.
94	3.7	Standard British (Ordnance QF) AA gun of WW2, static and mobile.
100	—	Japanese naval dual-purpose guns WW2 and standard current rocket size.
102	4	Coastal guns adapted for AA role by Britain and Italy.
105	—	Standard field artillery calibre. British, German and Japanese HAA guns, limited American use and static Russian use in this role. Air portable guns.
—	4.45	Ordnance QF guns of WW2 in British service. Captured examples in German service.
120	4.7	Japanese naval dual purpose guns and over 500 guns of this calibre built for AA use in America 1918-45.
125	—	Standard rocket size.
127	5.0	Japanese naval dual purpose guns WW2 and current Japanese rocket size.
128	—	German Flak 40 high performance static and rail-mounted HAA gun of WW2. Current rocket size (eg, Yugoslavia).
130	—	Current rocket size (eg, Czech).
—	5.25	Ordnance QF HAA gun introduced 1943.
135	—	Standard rocket size (eg, Bofors).

IX World absolute speed records, 1909-80

Date	Km/h	Mph	Pilot	Aircraft	Remarks, location, etc
May 20 '09	54.87	34.1	Paul Tissander	Wright Flyer	Pau, France
Aug 23 '09	69.7	43.3	Glenn Curtiss (USA)	Herring-Curtiss biplane	Reims, France
Aug 24 '09	74.3	46.1	Louis Blériot	Blériot XII/XI	On Aug 28 to 76-99 km/h
Apr 23 '10	77.6	48.2	Hubert Lantham	Antoinette monoplane	Nice, France
Jul 10 '10	106.5	66.1	Leon Morane	Blériot monoplane	Reims, France
Oct 29 '10	109.7	68.1	Alfred Le Blanc	Blériot monoplane	On Apr 12 '11 to 111.79 km/h
May 11 '11	119.7	74.3	Édouard de Nieuport	Nieuport	Chalons, France
Jun 12 '11	125.0	77.6	Alfred Le Blanc	Blériot monoplane	Pau, France
Jun 16 '11	130.0	80.7	Édouard de Nieuport	Nieuport	On Jun 21 to 133.11 km/h
Jan 13 '12	145.1	90.1	Jules Vedrines	Deperdussin monoplane	7 stages to 174.06 km/h
Jun 17 '13	179.8	111.7	Maurice Prévost	Deperdussin monoplane	3 stages to 203.81 km/h
Feb 7 '20	275.2	171.0	Sadi Lecointe	Nieuport-Delage 29	Villacoublay, France
Feb 28 '20	283.4	176.0	Jean Casale	Spad Herbemont	Villacoublay, France
Oct 9 '20	292.6	181.7	Baron de Romanet	Spad Herbemont	Buc, France
Oct 10 '20	296.9	184.4	Sadi Lecointe	Nieuport-Delage 29	On Oct 20 to 302.48 km/h
Nov 4 '20	309.0	192.0	Baron de Romanet	Spad Herbemont	Buc, France
Dec 12 '20	313.0	194.4	Sadi Lecointe	Nieuport-Delage 29	On Sep 21 '22 to 341 km/h
Oct 18 '22	358.7	222.8	Brig Gen W.A. Mitchell	Curtiss R-6	Detroit, Michigan
Feb 15 '23	374.9	233.0	Sadi Lecointe	Nieuport Delage	Istres, France in 3 stages
Mar 29 '23	380.7	236.4	Lt* R.L. Maughan	Curtiss R-6	Ohio, USA
Nov 2 '23	411.0	255.2	Lt H.J. Brow	Curtiss R2C-1	Mitchell Field, NY
Nov 4 '23	429.9	266.9	Lt A.J. Williams	Curtiss R2C-1	Mitchell Field, NY
Dec 11 '24	448.1	278.3	Adj Chef A. Bonnet	SIMB V-2	Istres, France
Nov 4 '27	479.2	297.5	Maj Mario de Bernardi	Macchi M-52	Venice, Italy
Mar 30 '28	512.7	318.4	Maj Mario de Bernardi	Macchi M-52*bis*	Venice, Italy
Sep 29 '31	654.9	406.7	Flt Lt G.H. Staniforth	Supermarine S6B	Calshot, England
Apr 10 '33	681.9	423.5	Lt F. Angello	Macchi-Castoldi 72	On Oct 23 to 709 km/h
Mar 30 '39	746.4	463.5	Flugkapitan H. Dietesle	Heinkel He100V-8	Oranienburg, Germany

Date	Km/h	Mph	Pilot	Aircraft	Remarks, location, etc
Apr 26 '39	754.9	468.8	Flugkapitan F. Wendel	Messerschmitt Bf109R	Augsburg, Germany
Nov 7 '45	975.6	605.8	Gp Capt H.J. Wilson	Gloster Meteor F4	Herne Bay, Kent
Sept 7 '46	990.8	615.3	Gp Capt E.M. Donaldson	Gloster Meteor F4	Rustington, Sussex
Jun 19 '47	1,003.6	623.3	Col Albert Boyd	Lockheed P-80R	Muroc, California
Aug 20 '47	1,030.9	640.2	Cdr T.F. Caldwell	Douglas D-558 Skystreak	Muroc, California
Aug 25 '47	1,047.3	650.4	Maj M.E. Carl, USMC		Muroc, California
Sep 15 '48	1,079.6	670.4	Maj R.L. Johnson, USAF	North American	Muroc, California
Nov 19 '52	1,123.9	697.9	Capt J. Slade Nash	F-86 A&D	Salton-Sea, California
Jul 16 '53	1,151.6	715.2	Lt Col W.F. Barnes	Sabres	USA
Sep 7 '53	1,170.7	727.0	Sqn Ldr Neville Duke	Hawker Hunter 3	Littlehampton, Sussex
Sep 25 '53	1,183.7	735.1	Lt Cdr M. Lithgow	Supermarine Swift 4	Libya, North Africa
Oct 3 '53	1,211.5	752.4	Lt Cdr J.B. Verdin	Douglas Skyray	Salton-Sea, California
Oct 29 '53	1,215.0	754.5	Lt Col F.K. Everest	YF-100A Super Sabre	Salton-Sea, California
Aug 20 '55	1,323.0	821.6	Col H.A. Hanes	F-100C Super Sabre	Edwards AFB, USA
Mar 10 '56	1,821.1	1,130.8	Lt Peter Twiss	Fairey FD2	Chichester, England
Dec 12 '57	1,943.0	1,206.6	Maj Adrian Drew	F-101A Voodoo	Los Angeles, USA
May 16 '58	2,259.2	1,403.0	Capt W.W. Irvin	F-104A Starfighter	Edwards AFB, USA
Oct 31 '59	2,387.5	1,483.0	Col G. Mosolov	Mikoyan E-66	Sidorovo, USSR
Dec 15 '59	2,455.2	1,524.7	Maj J.W. Rogers	F-106A Delta Dart	Edwards AFB, USA
Nov 22 '61	2,585.4	1,605.5	Lt Col R. Robinson	F4H-1 Phantom	Edwards AFB, USA
Jul 7 '62	2,681.5	1,665.2	Col R. Stephens	Lockheed YF-12A	Edwards AFB, USA
Jul 28 '76	3,529.6	2,191.9	Capt E.W. Joersz	Lockheed SR-71A	Beale AFB, California

*For the sake of brevity all ranks have been abbreviated in these tables.

X World absolute distance records, 1908-80

Date	Km	Miles	Pilot	Aircraft	Remarks, locations, etc
Jan 13 '08	1	0.6	Henry Farman	Voisin	To 2 km Mar 21 '08
Sep 17 '08	24.1	15.0	L. Delagrange	Voisin	In 3 stages, France
Dec 31 '08	124.7	77.44	Wilbur Wright	Wright	In 3 stages, France
Aug 25 '09	134.0	83.3	Louis Paulhan	Voisin	Betheny, France
Aug 26 '09	154.6	96.0	Hubert Lantham	Antoinette IV	Betheny, France

Jul 20 '10	392.7	244.0	Jan Olieslagers	Blériot	Mourmelon, France
Oct 28 '10	465.7	289.2	M. Tabuteau	Maurice Farman	Etampes, France
Dec 11 '10	515.9	320.37	G. Legagneux	Blériot	Pau, France
Dec 30 '10	584.7	363.1	M. Tabuteau	Maurice Farman	Buc, France
Jul 16 '11	625.0	388.1	Jan Olieslagers	Nieuport Monoplane	Belgium
Sep 1 '11	722.9	448.9	George Fourny	Maurice Farman	Buc, France
Dec 24 '11	740.3	460.0	A. Gobé	Nieuport Monoplane	Pau, France
Sep 11 '12	1,010	627.2	M. Fourny	Maurice Farman	Etampes, France
Oct 13 '13	1,021	634.0	A. Seguin	Henry Farman	Paris-Bordeaux return
Feb 3/4 '25	3,166	1,966	Arrachart and Lemaitre	Breguet 19	Etampes-Villa Cisneros
Jun 26/7 '26	4,305	2,673	L. & A. Arrachart	Potez 550	Le Bourget-Basra
Jul 14/5 '26	4,715	2,928	Girier & Dordilly	Breguet 19	Le Bourget-Omsk
Sep 31/1 '26	5,174	3,213	Châlle and Weiser	Breguet 19	Le Bourget-Banda Abbas
Oct 28/9 '26	5,396	3,351	Costes and Rignot	Breguet 19	Le Bourget-Jask
May 20/1 '27	5,809	3,607	Charles Lindbergh	Ryan Monoplane	New York-Paris
Jun 4/6 '27	6,294	3,909	Chamberlin and Levine	Bellanca WB2	New York-Germany
Jul 3/5 '28	7,188	4,464	Farrarin and Prete	Savoia S64	Rome-Brazil
Sep 27/9 '29	7,905	4,909	Costes and Bellonte	Breguet 19	Le Bourget-Moulart
Jul 28/30 '31	8,065	5,008	Boardman and Polando	Wright J6	Brooklyn, USA
Feb 6/8 '33	8,544	5,306	Gayford and Nicholetts	Fairey Monoplane	Cranwell-SW Africa
Aug 5/7 '33	9,104	5,654	Rossi and Codos	Blériot 110	New York-Rayak
Jul 13/5 '37	10,148	6,502	Col Gromov plus 2	ANT-25-1	Moscow-California
Nov 5/7 '38	11,520	7,154	R. Kellett (leader)	2 Vickers Wellesleys	Ismailia-Darwin
Nov 12 '45	12,739	7,911	Col Irving and crew	B-29 Superfortress	Japan-Washington
Oct 29/1 '46	18,082	11,229	Col Irvine and crew	Lockheed Neptune	Perth-Columbus
Jan 10/11 '62	20,168	12,524	Cdr T. Davies and crew	B-52H Stratofortress	Okinawa-Madrid

XI World absolute height record, 1909-80

Date	Metres	Feet	Pilot	Aircraft	Remarks, locations, etc
Aug 29 '09	155	508	Hubert Latham	Antoinette	Reims, France
Oct 18 '09	300	984	Comte C. de Lambert	Wright	Paris, France

Date	Metres	Feet	Pilot	Aircraft	Remarks, locations, etc
Jan 7 '10	1,000	3,280	Hubert Latham	Antoinette	In 2 stages, France
Jan 12 '10	1,209	3,966	Louis Paulhan	Henry Farman	Los Angeles, USA
Jun 14 '10	1,335	4,379	Walter Brooking	Wright biplane	Idianapolis, USA
Jul 7 '10	1,384	4,540	Hubert Latham	Antoinette	Reims, France
Jul 10 '10	1,900	6,232	Walter Brooking	Wright biplane	Atlantic City, USA
Aug 11 '10	2,012	6,599	J.A. Drexel	Blériot monoplane	Lanark, Scotland
Sep 3 '10	2,582	8,469	Leon Morane	Blériot monoplane	Deauville, France
Sep 8 '10	2,587	8,485	George Chavez	Blériot monoplane	Issy, France
Oct 1 '10	2,780	9,118	H. Wynmaler	Henry Farman	Mourmelon, France
Oct 23 '10	2,880	9,446	J.A. Drexel	Blériot monoplane	Philadelphia, USA
Oct 31 '10	2,960	9,709	Ralph Johnstone	Wright biplane	Belmont Park, NY
Dec 8 '10	3,100	10,168	G. Legagneux	Blériot monoplane	Pau, France
Jul 8 '11	3,177	10,421	M. Loridan	Henry Farman	Chalons, France
Aug 9 '11	3,190	10,463	Commandant Felix	Blériot monoplane	Etampes, France
Sep 6 '12	4,900	16,072	Roland Garros	Blériot XI	In 2 stages, France
Sep 17 '12	5,450	17,876	G. Legagneux	Morane monoplane	Corbeaulieu, France
Dec 11 '12	5,610	18,401	Roland Garros	Morane monoplane	Tunis, North Africa
Mar 11 '13	5,880	19,286	M. Perreyon	Blériot XI	Buc, France
Dec 28 '13	6,120	20,074	G. legagneux	Nieuport	St Raphael, France
Feb 27 '20	10,093	33,105	Maj R. Schroeder	Le Pere Lusac 11	Dayton, USA
Sep 18 '21	10,518	34,499	Lt J.A. MacReady	Le Pere Lusac 11	Dayton, USA
Oct 30 '23	11,145	36,556	Sadi Lecointe	Nieuport 40	2 stages, France
Jul 25 '27	11,710	38,409	Lt C.C. Champion	Wright F3W-1	Anacostia, USA
May 8 '29	11,930	39,130	Lt A. Soucek	Wright F3W-1	Anacostia, USA
May 26 '29	12,739	41,784	W. Neuenhofen	Junkers W34	Dessau, Germany
Jan 4 '30	13,157	43,155	Lt A. Soucek	Wright F3W-1	Anacostia, USA
Sep 16 '32	13,404	43,965	Capt C.F. Uwins	Vickers Vespa	Filton, nr Bristol
Sep 28 '33	13,661	44,808	G. Lemoine	Potez 506	Villacoublay, France
Apr 11 '34	14,433	47,340	Cdr R. Donati	Caproni 133	Rome, Italy
Aug 14 '36	14,843	48,685	G. Détré	Potez 506	Villacoublay, France
Sep 28 '36	15,223	49,931	Sqn Ldr F.R.D. Swain	Bristol 138	Farnborough, Hants

Jun 30 '37	16,440	53,923	Flt Lt M.J. Adam	Bristol 138A	Farnborough, Hants
Oct 22 '38	17,087	56,045	Lt Col M. Pezzi	Caproni 161 *bis*	Montecelio, Italy
Mar 23 '48	18,119	59,430	J. Cunningham	DH Vampire	Hatfield, Herts
Aug 29 '55	20,083	65,872	Wg Cdr W.F. Gibb	Canberra (WD952)	Filton, nr Bristol
Aug 28 '57	21,430	70,290	Michael Randrup	Canberra (WK163)	Luton, Bedfordshire
Apr 18 '58	23,449	76,913	Lt Cdr G. Watkins	F11F-1F Tiger	Edwards AFB, USA
May 2 '58	24,217	79,432	Roger Carpentier	SO 9050 Trident II	France
May 7 '58	27,811	91,220	Maj H.C. Johnson	F-104A Starfighter	Palmdale, USA
Jul 14 '59	28,852	94,635	Maj V. Ilyushin	Sukhoi T-431	USSR
Dec 6 '59	30,040	98,531	Cdr L.E. Flint	F4H-1 Phantom	Edwards AFB, USA
Dec 14 '59	31,513	103,363	Capt J.B. Jordan	F-104C Starfighter	Edwards AFB, USA
28 Apr '61	34,714	113,862	Col G. Mosolov	Mikoyan E-66A	USSR
Jul 25 '73	36,240	118,867	Col A. Fedotov	Mikoyan E-266	USSR
Aug 31 '77	37,650	123,492	Col A. Fedotov	Mikoyan E-266M	USSR

XII Schneider Trophy contest winners

Year	Nation	Pilot	Aircraft	Venue	Average speed (mph)	Remarks
1913	France	M. Prévost	Deperdussin	Monaco	45.75	US was runner-up
1914	UK	C.H. Pixton	Sopwith Tabloid	Monaco	86.75	Switzerland was runner-up
1920	Italy	Lt L. Bologna	Savoia S-12	Venice	107.2	No other contestant
1921	Italy	Lt Briganti	Macchi M7	Venice	117.86	Rest failed course
1922	Italy	Capt H.C. Biard	Supermarine Sea Lion III	Naples	145.70	Italy only contested
1923	USA	Lt D. Rittenhouse	Navy Curtiss R-3	Cowes	177.38	France and UK contested
1925	USA	Lt J. Doolittle	Army Curtiss R3C-2	Baltimore	232.57	Italy and UK contested
1926	Italy	Maj de Bernard	Macchi M-39	Hampton Rds	246.5	US and UK contested
1927	UK	Flt Lt S.N. Webster	Supermarine S5	Venice	281.65	Italy only contested
1929	UK	Flg Off Waghorn	Supermarine S6	Cowes	328.63	Italy only contested
1931	UK	Lt J.H. Boothman Flt Lt J.H. Boothman	Supermarine S6B	Solent	340.1	Not contested

Trophy won outright by Britain by 3 successive wins.

Index

A Search and Rescue Sikorsky S-58.

MAR NE
07